Alternatives to the Current Use of Nitrite in Foods

Part 2 of a 2-Part Study by the Committee on Nitrite and Alternative Curing Agents in Food

Assembly of Life Sciences

NATIONAL ACADEMY PRESS
Washington, D.C. 1982

NOTICE: The project that is the subject of this report was approved by the Governing Board of the National Research Council, whose members are drawn from the Councils of the National Academy of Sciences, the National Academy of Engineering, and the Institute of Medicine. The members of the Committee responsible for the report were chosen for their special competences and with regard for appropriate balance.

This report has been reviewed by a group other than the authors according to procedures approved by a Report Review Committee consisting of members of the National Academy of Sciences, the National Academy of Engineering, and the Institute of Medicine.

The National Research Council was established by the National Academy of Sciences in 1916 to associate the broad community of science and technology with the Academy's purposes of furthering knowledge and of advising the federal government. The Council operates in accordance with general policies determined by the Academy under the authority of its congressional charter of 1863, which establishes the Academy as a private, nonprofit, self-governing membership corporation. The Council has become the principal operating agency of both the National Academy of Sciences and the National Academy of Engineering in the conduct of their services to the government, the public, and the scientific and engineering communities. It is administered jointly by both Academies and the Institute of Medicine. The National Academy of Engineering and the Institute of Medicine were established in 1964 and 1970, respectively, under the charter of the National Academy of Sciences.

The work on which this publication is based was funded by the U.S. Department of Agriculture and the U.S. Food and Drug Administration under USDA Contract No. 53-3A94-0-18.

Library of Congress Catalog Card Number 82-81677

International Standard Book Number 0-309-03277-6

Available from

NATIONAL ACADEMY PRESS
2101 Constitution Ave., N.W.
Washington, D.C. 20418

Printed in the United States of America

COMMITTEE ON NITRITE AND ALTERNATIVE CURING AGENTS IN FOOD

SEYMOUR COHEN, Department of Pharmacological Sciences, State University of New York, Stony Brook, New York, Advisor

PHILIP E. HARTMAN, Department of Biology, Johns Hopkins University, Baltimore, Maryland, Advisor

PETER N. MAGEE, Fels Research Institute, Temple University School of Medicine, Philadelphia, Pennsylvania, Advisor

ROLF PREUSSMANN, Deutsches Krebsforschungszentrum, Institut für Toxikologie und Chemie, Heidelberg, Federal Republic of Germany

National Research Council Staff

SUSHMA PALMER, Project Director

ROY WIDDUS, Staff Officer

CATHERINE ST. HILAIRE, Staff Officer

LESLIE GRAYBILL, Research Associate

ROBERT HILTON, Research Associate

NORMAN GROSSBLATT, Editor

PREFACE

In September 1980, the U.S. Department of Agriculture and the Food and Drug Administration asked the National Academy of Sciences to examine the health effects of dietary nitrate and nitrite and to evaluate possible alternatives to the use of nitrite as a preservative in food. Accordingly, the Committee on Nitrite and Alternative Curing Agents in Food was established in the Assembly of Life Sciences of the National Research Council and given the task of reviewing scientific literature pertaining to these subjects and preparing two reports. In its first report, the Committee assessed the health risks associated with overall exposure to nitrate, nitrite, and *N*-nitroso compounds, placing emphasis on the risks associated with natural and added nitrate and nitrite in food and the utility of nitrite added to food. In this second report, the Committee reviews the status of research and future prospects for developing feasible alternatives to the use of nitrite as a preservative.

A special effort was made to ensure that the collective knowledge of the Committee encompassed all the types of expertise needed to conduct a study of this scope. The resulting multidisciplinary group includes the biomedical expertise that was needed to evaluate the toxicologic and carcinogenic significance of exposures to food additives and environmental chemicals, the metabolism and pharmacokinetics of such xenobiotic compounds, and the practicality, antimicrobial efficacy, and utility of food additives.

The search for information needed by the Committee went beyond a review of the scientific literature to include requests for information from scientists not on the Committee, federal agency officials, and consultants in the food industry and trade associations. Consultants and other persons were also invited to make oral presentations or to prepare papers for consideration by the Committee. In January 1981, a widely advertised public meeting was held in an attempt to ensure that all those wishing to contribute material to the Committee had the opportunity to do so.

The Committee recognized that the subject of its study was of great interest to the public, which is concerned about the safety of food; to the food industry, which must provide a safe and economical product while earning a fair commercial return; and to the regulatory agencies, which are responsible for monitoring production and distribution of food in order to protect public health. Having recognized these interests, the Committee directed its attention to the study of scientific data that were important in the development of public policy concerning the addition of nitrite to food.

To address its separate tasks, the Committee formed from among its members two closely interacting subgroups. Each was given primary responsibility for the initial analysis of the evidence pertaining to one of the tasks--the health effects of nitrate, nitrite, and *N*-nitroso compounds and the status of research on alternative curing agents. However, each subgroup contributed to the work of the other through discussions and shared writing efforts. The Committee as a whole reviewed each report, and resulting comments have been incorporated into the text.

The scientific questions addressed by the Committee were complex. Among those most pertinent to this second report were the following: For what purposes is nitrite added to foods, i.e., what are its effects? What is the evidence that alternatives to achieve these effects are needed? How do current food production and handling practices determine the properties that alternatives should have? Can the putative risk of nitrite use be ameliorated by means other than its replacement? Are there agents or processes that produce all or some of the effects of nitrite with equal efficacy? If so, are their effects on health understood? What research would most rapidly test the feasibility of using suggested alternatives in various products?

This report attempts to answer those questions. A summary of the Committee's findings, conclusions, and recommendations appears in the first chapter. Chapter 2 reviews some of the evidence presented in the Committee's first report, on the use of nitrite in cured meats and its effects, and outlines the Committee's approach in evaluating alternatives to nitrite. Chapter 3 presents general information on the preservation of cured meats, and Chapters 4 and 5 review the data available on antimicrobial alternatives. Chapters 6 and 7 evaluate alternatives for two other effects of nitrite--inhibition of lipid oxidation and development and maintenance of cured-meat color. In Chapter 8, the various means of inhibiting nitrosamine formation in cured meats are evaluated. Chapters 9 and 10 review the information on flavor and toxicity for the most promising alternatives identified by the Committee in Chapters 4 through 8. Finally, Chapter 11 outlines the Committee's suggestions with respect to long-term research designed to develop alternatives to nitrite.

Economic considerations were not addressed in this report, because the Committee believed that such considerations were not part of its charge.

The Committee is grateful to all who contributed to this report. It wishes especially to acknowledge the contributions of the following consultants, who provided valuable information and, at the request of the Committee, drafted manuscripts for review and use by the Committee:

- Robert Cassens, Muscle Biology Laboratory, University of Wisconsin, Madison, Wisconsin

- Melvin Eklund, Northwest and Alaska Fisheries Center/Utilization Research Division, U.S. Department of Commerce, National Oceanic and Atmospheric Administration/National Marine Fisheries Service, Seattle, Washington

- Harold L. Newmark, Hoffmann-La Roche, Incorporated, Nutley, New Jersey

- Glen R. Schmidt, Department of Animal Science, Colorado State University, Fort Collins, Colorado

- Bruce Tompkin, Research and Development Center, Swift and Company, Oak Brook, Illinois

Other organizations and individuals that kindly supplied requested information or advice to the Committee include the American Meat Institute; T. A. Roberts, Meat Research Institute, Bristol, United Kingdom; L. Leistner, Institute for Microbiology, Toxicology and Histology, Bundesanstalt für Fleischforschung, Kulmbach, Federal Republic of Germany; J. N. Sofos, Colorado State University; M. D. Pierson, Virginia Polytechnic Institute, Blacksburg; M. C. Robach, Monsanto Chemical Company, St. Louis, Missouri; T. Tsuzuki, Ueno Fine Chemicals Co., New York; J. J. Kabara, Michigan State University, East Lansing; N. Weinshenker, Dynapol Inc., Palo Alto, California; and the Environmental Mutagen Information Center and the Environmental Teratology Information Center, Oak Ridge National Laboratory, Tennessee.

The Committee is also grateful to all those persons who presented useful data and participated in the public meeting held on January 22, 1981.

Thanks are also due to the Board on Toxicology and Environmental Health Hazards and the Food and Nutrition Board, both units of the National Research Council's Assembly of Life Sciences, and to James Kirk, Philip Landrigan, Robert Neal, and Steven Tannenbaum, who provided advice and assistance in planning the study.

The Committee acknowledges the kindness of Ronald E. Engel, Thomas P. Grumbly, and Donald L. Houston of the U.S. Department of Agriculture and Albert C. Kolbye of the U.S. Food and Drug Administration and their colleagues, who provided access to information and shared their insights.

The staff of the Assembly of Life Sciences--Sushma Palmer, Roy Widdus, Catherine St. Hilaire, Leslie Graybill, Robert Hilton, and Norman Grossblatt--provided the extensive support and assistance that were essential to the completion of the report. The Committee is deeply indebted to them for their work. The dedicated secretarial services of Susan G. Barron, Juanita E. Maclin, Roberta A. Piorsko, and Chipella Sumby-Reap also contributed to the timely completion of this report.

MACLYN McCARTY
Chairman
Committee on Nitrite and
Alternative Curing Agents in Food

CONTENTS

CHAPTER 1

SUMMARY, CONCLUSIONS, AND RECOMMENDATIONS

Page

CHAPTER 1

SUMMARY, CONCLUSIONS, AND RECOMMENDATIONS

Nitrate, and probably nitrite, have been used for centuries in curing salts to preserve meat. Their initial use was inadvertent--nitrate (as saltpeter) was added as a contaminant of the salt used to cure meat. By the late 1800s, however, the value of saltpeter in promoting cured-meat color was recognized, and its addition to meats was recommended. It was later discovered that nitrate yields nitrite, which is responsible for cured color. Nitrite inhibits the growth of some bacteria that cause meat to spoil and of *Clostridium botulinum*, the microorganism that causes botulism.

Currently, nitrite (as sodium nitrite) is added to a wide variety of cured meats, and nitrate is used primarily in the curing of foods, such as fermented sausages and dry-cured meats, that require long production times. In these food items, nitrate serves as a reservoir for the production of nitrite, through the bacterial reduction of nitrate. Sodium nitrite is added to most bacon at 120 mg/kg, to other pickle-cured products at 200 mg/kg, and to most comminuted (chopped or ground) cured meats at 156 mg/kg. The addition of nitrate and nitrite to meats is regulated by the U.S. Department of Agriculture (USDA). The use of these agents is also permitted in some types of smoked fish. This use of nitrate and nitrite is regulated by the Food and Drug Administration (FDA).

In recent years, the finding that nitrite can interact with various amino compounds to produce carcinogenic *N*-nitroso compounds, such as nitrosamines, has caused considerable concern over possible adverse health effects and has resulted in a search for safer preservatives. In its first report, the Committee on Nitrite and Alternative Curing Agents in Food evaluated the health effects of nitrate, nitrite, and *N*-nitroso compounds. Because nitrate and nitrite contribute to human exposure to *N*-nitroso compounds, which are carcinogenic in laboratory animals and may be carcinogenic in humans, the Committee recommended that exposure of humans to these agents from all sources be reduced.

Humans are exposed to nitrate and nitrite from a variety of sources. For example, for the average U.S. citizen, most nitrite comes from cured meats (∿39%) and baked goods and cereals (∿34%). However, 87% of nitrate ingested comes from vegetables and when the conversion of nitrate to nitrite in the human body is considered, most of the nitrite to which the average U.S. citizen is exposed actually comes from vegetables (∿72%), and less than 10% comes from cured meats. Although the contribution of cured meats to the total body burden of nitrite was estimated by the Committee to be less than 10%, the Committee recommended that the amount of nitrite added to cured meats be reduced "to the extent that protection against botulism is not compromised." The Committee also recommended that the search for alternatives to nitrite in cured meats be continued and that no new agent or combination of agents be substituted for nitrite until adequate testing has ensured that it does not present a greater hazard to human health.

The Committee has now examined the status of research on proposed alternatives to the use of nitrite in cured meats to determine whether suitable agents are available. In addition, strategies for long-term research on alternatives to nitrite were developed.

Although nitrite has been used as a preservative for many years, its specific effects in cured meats have been investigated only relatively recently. Studies have revealed that, in general, nitrite has antimicrobial, antioxidant, and sensory (color and flavor) properties and that the necessity for these effects of nitrite varies among different cured-meat products.

The need for nitrite or an alternative as an antimicrobial agent depends on the degree of contamination of a product, the ability of intrinsic product characteristics to inhibit microbial proliferation, and the likelihood that the product will be exposed to temperatures that permit microbial growth. On the basis of data currently available, shelf-stable canned cured meats and perishable canned meats are the products most likely to become toxic or spoiled if nitrite addition is reduced or omitted without an effective replacement. Dry and semidry fermented sausages, dry-cured cuts, some bacons (e.g., those containing fermentable carbohydrate), and commercially sterile products are less likely to become toxic or spoiled in the absence of nitrite or other substances with antimicrobial properties.

Proper refrigeration can reduce the risk of botulism from perishable items, and, hence, the need for nitrite in these products. This method could be made more reliable if consumers and others who handle these products were better educated concerning the importance of proper refrigeration in the safety of these items.

Nitrite fixes the color in most cured meats; however, the data indicate that the concentration of nitrite required to fix color varies. Sodium nitrite at 40-50 mg/kg is sufficient to produce adequately stable cured-meat color in most products. Data on the contribution of nitrite to the flavor of products and on the amount of nitrite needed to inhibit lipid oxidation in various cured meats are incomplete; thus, the necessity for nitrite in producing these effects in different products and the concentrations required are unclear.

Alternatives to the conventional use of nitrite are of two general categories--agents or treatments that serve as partial or complete replacements for nitrite, and agents that block the formation of nitrosamines in products containing conventional concentrations of nitrite. The Committee assessed the ability of alternatives in the first category to substitute completely or partially for nitrite in each of its effects. In general, because most alternatives tested produce only one of the effects of nitrite (usually antimicrobial), combinations of compounds are often necessary to achieve all the actions exerted by nitrite. In many cases, the combination includes small amounts of nitrite sufficient to produce the cured color (40-50 mg/kg). In such cases, the Committee evaluated the combination rather than each individual agent. The Committee evaluated agents in the second category for their ability to inhibit the formation of nitrosamines in bacon.

For each of the alternatives, or combinations of alternatives, shown to be effective in one or more cured-meat products, the Committee reviewed data on sensory characteristics to determine whether the products so treated would be acceptable to consumers, as well as data on acute and chronic toxicity to determine whether the proposed alternative could present a hazard to human health.

On the basis of its overall review, the Committee found the following proposed alternatives to be most promising at present: the combination of ascorbate, α-tocopherol, and nitrite; irradiation (with or without nitrite); lactic-acid-producing organisms (with or without nitrite); potassium sorbate with low concentrations of nitrite; sodium hypophosphite (with or without nitrite); and several fumarate esters.

All the above-mentioned alternatives, with the exception of ascorbate and α-tocopherol, serve as partial or complete replacements for the antimicrobial effects of nitrite. Ascorbate and α-tocopherol function as inhibitors of nitrosamine formation.

STATUS OF PROPOSED ALTERNATIVES

Ascorbate and α-Tocopherol

Ascorbate is an excellent inhibitor of nitrosamine formation in nonfood systems, especially at low pH. It is also effective in a variety of cured meats, including frankfurters, ham, and bacon. The USDA currently requires that sodium ascorbate or sodium erythorbate at 550 mg/kg be used to reduce the amount of nitrosamines formed in bacon--the cured-meat product in which the highest concentrations of nitrosamines are found. However, because nitrosamines are formed during cooking in the lipid portion of bacon, ascorbate, with only slight solubility in lipids, is not completely successful.

α-Tocopherol also inhibits the formation of nitrosamines, and it is soluble in lipids. Evidence that α-tocopherol is effective in meat systems comes primarily from studies in pickle-cured bacon, in which significant reductions in nitrosamine formation have been reported. Because α-tocopherol is insoluble in aqueous cure mixtures, special methods, such as the incorporation of polysorbate emulsifiers in the cure, are required to ensure adequate distribution in the product. α-Tocopherol can also be applied to the surface of the product by spraying or dipping--frying disperses it throughout the bacon. α-Tocopherol is effective in bacon at a concentration of 500 mg/kg.

The combination of ascorbate and α-tocopherol (ascorbate or erythorbate at 550 mg/kg and α-tocopherol at 500 mg/kg) has been shown to be effective in inhibiting nitrosation in bacon and does not interfere with the antibotulinal activity of nitrite in bacon when tested shortly after processing (1-10 days). However, the effect of the combination on antibotulinal activity has not been tested after longer periods of storage. In addition, although ascorbate and α-tocopherol have been tested in toxicity assays separately and have been shown to have no adverse health effects, there have been no toxicity tests of the combination of ascorbate, α-tocopherol, and nitrite.

Irradiation

The use of ionizing radiation, such as gamma irradiation, is an established method of microbial inactivation and has been studied as a method for preserving many foods. It has been found effective against *C. botulinum* in corned beef and has been tested in combination with low concentrations of sodium nitrite (25-40 mg/kg) in ham and bacon. Thus, irradiation may either substitute for the antimicrobial effects of nitrite or be used to reduce the concentration of sodium nitrite required for antimicrobial activity. Because irradiation would not prevent lipid oxidation or impart cured color or flavor to a product, it would probably be used most commonly in combination with low concentrations (40-50 mg/kg) of sodium nitrite. Bacon treated with a combination of irradiation and low concentrations of sodium nitrite has been evaluated by sensory panels for flavor and has been found to be acceptable. However, although the toxicity of irradiated meat and fish has been investigated in several studies, the Committee was unable to find any published reports on the toxicity of irradiated cured meats containing low concentrations of nitrite.

Lactic-Acid-Producing Organisms

Lowering the pH of cured-meat products can reduce spoilage and inhibit proliferation of microbial pathogens. The incorporation of lactic-acid-producing bacteria and fermentable carbohydrates in cured-meat formulations is currently permitted to reduce product pH;

however, nitrite, at conventional concentrations, is also included. Addition of lactic-acid-producing bacteria and fermentable carbohydrate has also been tested as a possible replacement for nitrite in bacon, a perishable (normally refrigerated) product in which acidulation would be activated by temperature abuse stimulating the metabolic (fermentative) activity of the lactic-acid-producing bacteria and their proliferation. Excellent protection against the formation of botulinum toxin was achieved in bacon with the addition of bacteria and sucrose with or without nitrite at 40 or 120 mg/kg. Bacon containing the combination of lactic-acid-producing bacteria and low concentrations of nitrite has been judged by sensory evaluation panels to have acceptable flavor.

Potassium Sorbate

The combination of potassium sorbate (2,600 mg/kg) and sodium nitrite (40-80 mg/kg) has been tested in bacon produced under a variety of commercial processing conditions and has antibotulinal activity equal to that of sodium nitrite at 120 mg/kg. The combination of sorbate and low nitrite is also active against *C. botulinum* in frankfurter emulsions and is effective against salmonellae and *Staphylococcus aureus*. Bacon containing this combination has been judged by a sensory evaluation panel to have acceptable flavor and color. Although sorbate alone is used in a variety of other foods and does not cause adverse effects in toxicity tests conducted in animals, several reaction products formed by the combination of sorbate and nitrite have recently been found to be mutagenic, and further testing of these reaction products in mammalian cell systems and in animals may be necessary to provide more definitive information on the toxicity of this combination.

Sodium Hypophosphite

Sodium hypophosphite, alone at 3,000 mg/kg or at 1,000 or 3,000 mg/kg in combination with sodium nitrite at 40 mg/kg, has antibotulinal activity in bacon at least equal to that of conventional nitrite use. This finding, however, is based on only two relatively small scale commercial plant studies. In addition, toxicity tests have not been conducted on sodium hypophosphite or the combination of sodium hypophosphite and nitrite. Sensory evaluation of bacon containing sodium hypophosphite, alone or in combination with low concentrations of nitrite, has been conducted, and both types of product have acceptable flavor.

Fumarate Esters

In a small-scale study of bacon produced under simulated commercial conditions, monomethylfumarate or monoethylfumarate at 1,250 mg/kg was as effective as sodium nitrite at 120 mg/kg. Sensory evaluation revealed that the methylfumarate-treated bacon was indistinguishable from nitrite-treated bacon. Toxicity testing on the fumarate esters has not been performed.

Conclusions and Recommendations

The most promising alternative to the conventional use of nitrite in bacon is the addition of α-tocopherol at 500 mg/kg in combination with ascorbate at 550 mg/kg and sodium nitrite at 120 mg/kg. This approach significantly reduces the formation of nitrosamines in bacon, and the Committee believes that it is unlikely to cause adverse health effects in humans when the above concentrations of α-tocopherol and ascorbate are used. However, the Committee recommends that mutagenicity and, if necessary, animal toxicity tests of the combination of ascorbate, α-tocopherol, and sodium nitrite be conducted to ensure that no reaction products are formed that could affect human health adversely. In addition, the Committee recommends that the effect of the combination of ascorbate and α-tocopherol on the antibotulinal activity of nitrite in bacon and other cured-meat products be evaluated at various times after processing to ensure that this activity of nitrite is not compromised.

Irradiation, alone and in combination with low concentrations of nitrite, has been shown to be effective against *C. botulinum* when tested in several cured-meat products. These findings need to be confirmed and extended to other products processed under a variety of commercial conditions. In addition, toxicity tests of the combination of irradiation and low concentrations of nitrite should be conducted.

Lactic-acid-producing bacteria, sodium hypophosphite, and several fumarate esters have each been shown to be effective in bacon against the formation of botulinum toxin; however, further tests of their effectiveness in other cured meats have not been conducted, nor has testing been confirmed under a variety of commercial conditions. The Committee recommends that such tests be performed and that the necessary toxicity testing be undertaken, if necessary.

Potassium sorbate in combination with low concentrations of nitrite (40-80 mg/kg) has been shown to be an effective antimicrobial agent in bacon and frankfurters produced under commercial conditions. Bacon containing this combination had acceptable flavor. The Committee recommends that tests for antimicrobial activity be conducted with this combination in other products made under commercial conditions. It also recommends that the question of the toxicity of mutagenic nitrite-sorbate reaction products be resolved.

RESEARCH RECOMMENDATIONS

The Committee offers a number of general research recommendations concerning future evaluations of the need for nitrite in various products, as well as long-term research strategies for developing additional alternatives.

Research on the Need for Nitrite in Various Products

Because the need for nitrite varies among cured-meat products (and, hence, the need for an alternative would also vary), the Committee recommends that the following research be conducted:

1. Because of the absence of data on the frequency of *C. botulinum* contamination in raw and cured meats, surveys should be conducted with serial sampling techniques to determine the extent of contamination. The effect of different production and geographic conditions on contamination frequency should be determined.

2. The relative frequency of temperature abuse in various classes of perishable cured products is unknown. Studies should be undertaken to determine this frequency and to assess the impact of public education on the frequency of temperature abuse.

3. More information is needed on the role of the various intrinsic factors (other than nitrite) in cured meats that contribute to the control of microbial proliferation. The interaction of factors that control pathogens and spoilage in different commercial products should be investigated (a list of the factors to be investigated is given in Chapter 3) to determine whether manipulation of those factors could reduce the need for nitrite in such products.

4. The need for antimicrobial agents in various cured meats should be reassessed periodically, especially after changes in production and handling practices.

5. The ability of nitrite to inhibit lipid oxidation and the concentration of nitrite required for this effect should be determined in a variety of cured meats.

6. The contribution of nitrite to the production of flavor in various cured meats should be further investigated (Chapters 7 and 9).

7. Although the role of nitrite in producing cured-meat color in various products is well established and the concentrations required for this effect are known, the strength and mutability of consumer preference for the traditional color of cured meats is less clear and should be investigated to determine whether there is a need for this particular effect of nitrite.

8. The Committee concluded in its first report that:

> Special considerations are relevant to the use of nitrate and nitrite in fish products. Most important among these are the higher frequency of contamination of fish with *C. botulinum* spores and the fact that the most common contaminating strains are able to grow at lower temperatures.

Additionally, there may be variation in the influence of alternatives on the sensory characteristics of products that would affect their acceptability. Hence, research should be conducted on alternatives that are applicable to fish products.

Long-Term Research Strategies for Developing Alternatives to Nitrite

The major effect of nitrite for which alternatives are being sought is its antimicrobial activity--especially its inhibition of spore-forming bacteria, particularly *C. botulinum*.

The most rational long-term approach to developing alternatives capable of eliminating the adverse effects of *C. botulinum* would be to seek agents that inhibit metabolic processes that are essential to this microorganism and other spore-formers, but are absent in humans. To accomplish this, more information on the physiology of spore-forming microorganisms is needed. Research emphasis should be on factors that affect dormancy, injury, activation, germination, and particularly outgrowth of spores, especially those of *C. botulinum*. Information on essential processes in the spore-to-cell transition could then be used to develop specific inhibitor(s) of the transition.

The research necessary to obtain this knowledge and to design and develop selective antimicrobial agents is likely to be laborious and expensive. Although such research would not be guaranteed to lead to the development of alternatives to nitrite in the near future, the Committee regards this approach as a necessary investment in the search for alternatives to nitrite. Such alternatives not only would reduce the risks associated with consumption of cured meats that contain nitrite, but also could be used in other food products in which spore-forming microorganisms currently pose the risks of botulism or spoilage.

CHAPTER 2

EVALUATING ALTERNATIVES TO NITRITE

Page

CHAPTER 2

EVALUATING ALTERNATIVES TO NITRITE

Salting of meat and fish to prevent them from spoiling or causing illness is an ancient practice. The salt used in these early curing procedures probably contained impurities, including nitrate, which can be reduced to yield nitrite. Over many centuries, the various effects attributable to the ingredients of curing salt became recognized, and processes were modified to optimize production of those effects. Color fixation was the first effect specifically attributed to nitrite. It was later recognized that nitrite had antimicrobial and thus, health protective functions, such as delaying the formation of *Clostridium botulinum* toxin. More recently, the antioxidant role of nitrite in inhibiting rancidity (resulting from lipid oxidation) has been recognized, and its contributions to the flavor of cured products, investigated.

Over the last two decades, however, concerns have arisen that both deliberately added and naturally occurring nitrate and nitrite in foods might have adverse long-term consequences for human health. In September 1980, the U.S. Department of Agriculture and the Food and Drug Administration asked the National Academy of Sciences to conduct a study of issues related to these concerns.

In its first report, the Committee on Nitrite and Alternative Curing Agents in Food, which was established by the National Academy of Sciences, described the use of nitrate and nitrite, the effects attributable to these compounds, and the potential health risks associated with exposure to them (National Academy of Sciences, 1981). The Committee recognized that nitrite under some conditions may react with nitrosatable substrates, such as amines and amides, to produce *N*-nitroso compounds, such as nitrosamines, many of which have been shown to be carcinogenic in laboratory animals. Nitrate can be reduced to nitrite and thus may indirectly contribute to *N*-nitroso compound formation.

The Committee concluded that reduction in exposure to nitrate and nitrite might lead to a reduction in risk. It recommended that the various exposures of humans to these compounds be reduced, because of their potential for contributing to exposures to *N*-nitroso compounds. Regarding the addition of nitrate and nitrite to foods, it recommended that the use of nitrate be discontinued in all but a few cured products, that the amount of nitrite be reduced to the extent that protection against botulism is not compromised, and that the search for alternatives to nitrite be continued.

In this report, the Committee has evaluated alternatives to nitrite by examining the efficacy of three general approaches to reducing the risks associated with the addition of nitrite to cured meats:

- Reduction in the amount of nitrite added.

- Substitution of another agent or treatment for nitrite.

- Use of nitrite in combination with an agent that inhibits the formation of N-nitroso compounds.

Because a thorough understanding of the current uses of nitrite and its role in various products is necessary for proper evaluation of alternatives to nitrite, the first two sections of this chapter review some of the conclusions of the Committee presented in its first report on the uses of nitrite and the most important effects of this agent in cured meats. The latter part of the chapter outlines the general strategy adopted by the Committee for the assessment of alternatives to nitrite, as well as a brief discussion of the types of alternatives considered by the Committee.

USE OF NITRATE AND NITRITE IN THE UNITED STATES

In the United States, nitrate and nitrite are used predominantly in curing red meats and poultry. Nitrate, if added, serves mostly as a reservoir from which nitrite is derived. Nitrate and nitrite are permitted in some kinds of smoked fish. In some countries, particularly European, nitrate is used in the production of cheese to prevent swelling; that practice is not permitted in the United States, so alternatives to nitrate for this purpose, which have been discussed elsewhere (Gray *et al.*, 1979), are not addressed here. Because of the large volumes of cured meats and poultry and the relatively small volume of nitrite-containing fish products, this report focuses on red meats and poultry and deals with fish products only where important special considerations arise.

Annual Production of Foods Containing Added Nitrate or Nitrite

The volume of production of cured-meat products to which nitrite is added and the general methods of addition are shown in Table 2-1. These categories and the methods of nitrite addition are explained more fully below. As the table illustrates, nearly 4 billion kilograms of cured meats were processed with added nitrite in the United States in 1979. This amounts to approximately 25% of all meat produced and includes a much higher percentage of some meats, such as pork. The use of nitrate in products has been decreasing (Binkerd and Kolari, 1975; Cerveny, 1980; Sofos and Busta, 1980), but it is still added to many products (Table 2-2). Most large processors have reportedly stopped adding nitrate to cured meats except fermented sausages and dry-cured cuts (B. Tompkin, Swift and Co., personal

TABLE 2-1

Meat Products Processed with Added Nitrite Under Federal Inspection in the United States During 1979[a]

Product	U.S. Production, billions of kilograms	Sodium or Potassium Nitrite Added,[b] mg/kg	Most Probable Method of Adding Nitrite
Cured beef	0.13	200	Injection
Ham, not canned	0.83	200[c]	Injection
Bacon	0.76	120	Injection
Sausage:			
Semidry and dry	0.15	156	In cure salt
Frankfurters	0.68	156	In cure salt
Bologna	0.37	156	In sure salt
Loaves, cured meat	0.05	156	In cure salt
Loaves, mixed meat and nonmeat ingredients	0.09	156	In cure salt
Liver	0.06	156	In cure salt
Other cooked items	0.38	156	In cure salt
Canned Meats:			
Hams	0.13	200	Injection
Luncheon meats	0.13	156	In cure salt
Hash	0.04	156	In cure salt
Vienna sausage	0.05	156	In cure salt
Miscellaneous	0.05	156	In cure salt

[a]Data from American Meat Institute, 1980, and U.S. Department of Agriculture, 1980. These numbers are subject to adjustment for the weight of nonmeat ingredients, weight loss during cooking, and production in other than federally inspected plants.

[b]Amount of nitrite added to meat products is based on amount of meat in the formulation. Thus, as extenders and other nonmeat ingredients are increased, nitrite added to total product is decreased. Most commonly, sodium nitrite is the nitrite salt used.

[c]This degree of nitrite addition does not apply to country-style (dry-cured) hams.

TABLE 2-2

Products Processed with Added Nitrate[a]

PORK

Bacon (brown-and-serve, Canadian-style, country-cured, country-style, jowl, pancetta rolled, pork shoulder); ears; fatback; hocks, jowls, loin; roll; shoulder butt; shoulder picnic; hams (canned, chopped, chopped loaf, boneless, country-style, cured, prosciutto, roll, semi-boneless, shanks, sliced, sliced boneless, Westphalian); smoked, chopped; spareribs

POULTRY

Cured parts, carcasses, products; sandwich; smoked; uncooked and stuffed; cured turkey ham

SAUSAGES

Cooked; cooked and smoked; dry; for pizza; Italian (cooked); Italian (cooked, cured); liver; New England brand; Polish (bratwurst); pork; semidry; smoked; smoked, country; summer; summer (cooked); summer (dry)

BEEF

Bologna (beef, garlic, Lebanon); breakfast beef; corned beef; corned beef brisket; corned beef hash; creamed, chipped; cured; frankfurters; in brine; jerky; meat bar; patties; roll, cooked; corned; smoked; sticks; tenderloin steaks; tongue, cured and smoked

SALAMI

Cooked; cotto; dry, hard; Genoa; German (dry); Italian (dry, hard)

OTHER PRODUCTS

Beerwurst; braunschweiger; bratwurst; cannelloni or tortellini; cappicola; cervelat; chorizo, dry; freizzes; galentini, head-cheese, sailcicca with cheese; Holsteinen; kielbassa; knockwurst; landjager; linguisa; cured meat loaves (beef, ham and pepperoni, pickle and pimento); longaniza; meats (e.g. poultry in sauces, and soy protein product, casseroles, cured mixes, cured patties, dehydrated, dressing with meat and poultry, in a blanket, luncheon, macaroni with meat, meat and gravy, omelets with meat or poultry or other components, or poultry, vegetables in gravy, pickled, salads, soups with meat, spreads); Milano; mortadella (cooked); pastels; pastrami; special items (burritos, corn dogs, crepes, enchiladas, hors d'oeuvres, jellied products, lima beans smoked--pork, ham, bacon in sauce, pate, pizza, quiche products, sauerkraut products, veal cordon bleu)

---

[a]Products were identified from new labels approved by the U.S. Department of Agriculture from 1979 to 1981 on which nitrate was listed as an ingredient. Data from U.S. Department of Agriculture, 1981, personal communication.

communication, 1981), and it is used mostly by small producers.[1] Nitrite is also permitted in smoked fish. Total U.S. production of commercially smoked fish was 9.6 million kilograms in 1979 (National Oceanic and Atmospheric Administration, 1980). Products in which nitrite was allowed constituted 5.8 million kilograms of this total.

Types of Products Containing Added Nitrite

It is difficult to describe the current U.S. use of nitrite (and nitrate) in cured red meats and poultry accurately, because processing, packaging, and distribution techniques are constantly changing. The variety of products to which nitrate and nitrite are added is large, procedures are diverse, and meat-processing facilities vary greatly in size and in the sophistication of their quality-control procedures. Many practices used widely in industry are not described in textbooks or examined in published research papers.

It is also difficult to generalize about product classes. One can define general categories into which most products will fit, but some products have characteristics, such as intermediate water activity,[2] that render them difficult to categorize. Many of the characteristics of finished products (such as water activity and pH) may result from traditional practices, rather than from procedures specifically designed to reach target values. The use of "least-cost" formulation methods may mean that the composition of a product varies even over relatively small periods, e.g., chicken meat is increasingly being incorporated into traditionally red meat products as a source of protein. Raw materials and methods of production, such as fermentation, may also vary among batches. Specific information on these variations is often not available.

Thus, any attempt at concise description of the current use of nitrite and nitrate runs the risk of oversimplifying a complex, varied, and ever-changing situation. The following description of

---

[1]Canadian manufacturing and handling practices are generally similar to those in the United States. Also, the climate in some regions of those countries is nearly identical. However, the use of nitrate in Canadian meat products has been forbidden, with some exceptions, for several years (Health and Welfare Canada, 1975).

[2]The water activity, a_w, of a food is defined as the ratio of the vapor pressure of water in the food (p) to that of pure water (p_o) at the same temperature, i.e., $a_w = p/p_o$. The growth of some organisms can be inhibited by lowering the water activity of a product.

the use of nitrite (and nitrate) in cured meats and poultry should therefore be regarded as general; in practice, the uses of these compounds are subject to much variation.

Cured-meat products can be divided into various categories on the basis of the extent to which they are heated (if at all) during production, of how they are cured, and of their water activity. Further subdivision into comminuted and noncomminuted (primal or "chunk") products is possible. The general categories of meat products are briefly described below to facilitate discussion of their susceptibility to microbial hazard and spoilage and of the role of nitrite in different products. In the following discussion, the term "processing" indicates manipulation, such as curing or comminution. The term "thermal processing" encompasses heating processes that control *C. botulinum*, but excludes milder heating, such as pasteurization and smoking.

- Raw, Cured Products with High Water Activity (>0.93): This category includes the form of corned beef that is packaged raw in free pickling solution (Price and Schweigert, 1971, p. 465). Some raw, cured products, such as bacon, are subjected to some smoking and mild heating during production (Kramlich *et al*., 1973, p. 228).

- Raw, Cured Products with Low Water Activity (<0.93): Scotch, prosciutto, Westphalian, and country ham, dry-cured bacon, and dried sausages may be cold-smoked and not heated appreciably during processing. Other meats in this category, such as some sausages, may receive some mild heating if smoked (Kramlich *et al*., 1973, p. 228; Nitrite Safety Council, 1980). These products would be sold as raw, cured products with a water activity that is low because of drying and salt addition. Many dried meats are produced with added nitrite, including dried beef, jerky, dry-cured bacon, dry-cured ham, and many dried sausages. Many of these depend on relatively high salt content, and consequently low water activity, for preservation at room temperature (Komarik *et al*., 1974).

- Cooked, Cured Products: This is by far the largest category of meats made with added nitrite (Price and Schweigert, 1971, p. 485).

Pasteurized products heated to a center temperature of 65-75°C include hams in casings and in cans, frankfurters, bologna, liver sausage, meat loaves, nonspecific loaves, and some roll products. Some pasteurized meats are sliced and packaged after heating; others are heated in their final containers. Bacon is classified as a raw, cured product, because it is generally not heated sufficiently to pasteurize it; but, with respect to its microbial profile, it is more akin to cooked, cured products.

A second class of cooked, cured meats includes the so-called *"shelf-stable" products*, such as canned luncheon meats and prefried canned bacon (Cerveny, 1980). These items are normally heated to a center temperature of 95-112°C, which alone is not sufficient to kill spores of _C. botulinum_ or other organisms, but in conjunction with other factors, such as the presence of nitrite, can delay their outgrowth. Many of these products contain not only meat, but also many other ingredients, such as cereal and soy proteins. This class also includes canned hams and shoulders.

Some cooked products--such as corned beef hash, deviled ham, meat spreads, and Vienna sausages--are defined as "_commercially sterile_," i.e., free of pathogens, as well as microorganisms capable of growing under normal nonrefrigerated storage conditions. They receive thermal processing of at least 2.78 min at 121°C.

Methods of Adding Nitrite to Meats

Effective preservation of meat frequently depends on uniform distribution of the active agent(s) throughout the product. Various methods of applying nitrate and nitrite have developed. In some cases, such as dry-curing (described below), the method of cure addition plays a large part in defining the ultimate sensory characteristics of a product.

Nitrite is added to meat products as a nitrite salt, usually sodium nitrite, as a nitrite-containing cure salt, or as a solution of nitrite and other ingredients referred to as "pickle." The method of addition may affect the uniformity of distribution of nitrite and thus the minimal concentration needed to achieve consistent results, e.g., in color development. Amounts of nitrite added to some types of meat products and methods of addition are listed in Table 2-1.

Curing pickle typically contains water, sodium chloride, sugar, phosphate, ascorbate (or isoascorbate), and sodium nitrite (Kramlich _et al_., 1973, p. 40). Curing pickle is used most frequently with intact meats that are not ground. In the case of intact, or primal, products that are made from portions of meat weighing more than about 100-200 g, the pickle is usually injected. With boneless products and sometimes with bone-in products, multineedle injectors are commonly used (Kramlich _et al_., 1973, p. 58). With properly adjusted multineedle injection, distribution of the nitrite-containing pickle can be reasonably uniform. In addition, some boneless hams and cured-beef products are subjected to mechanical treatment--tumbling or massaging--after pickle injection to distribute the pickle more uniformly.

Nitrite is generally added to ground or chopped (comminuted) meats in cure salt or as sodium nitrite during the blending of the meat. This usually results in very uniform distribution of the nitrite. In sausage manufacture, the meat is ground again and blended or chopped to the desired consistency. Coarsely comminuted products are used for dry and semidry sausages and some loaf products. Very finely comminuted emulsions are used for a large variety of skinless frankfurters, bologna, and loaves.

Finally, a number of products have nitrite or nitrate (or both) applied as part of a dry rub (Kramlich *et al*., p. 52; Price and Schweigert, 1971, p. 463). The rub usually contains sodium chloride, sugar, and sodium nitrite or nitrate (or both), which are blended and then mechanically or manually applied to hams, bellies, or beef.

Handling of Cured Products after Processing

The time to consumption of cured products may vary from less than a day to more than a year after production. For example, products from local and regional processors may be consumed within 2 wk. However, most cured-meat products consumed in the United States are distributed through complex distribution chains. A product from a national packer may have to pass through the producer's warehouse, a broker, a retail warehouse, and a retail store. Perishable products must remain wholesome and acceptable for 1-2 mo. Other products, such as pasteurized canned hams, may be consumed up to a year after production. During this time, temperature abuse[1] may occur in the production plant (e.g., on the loading dock), in transit during distribution, at a warehouse, at a retail store (e.g., at the top of a display case) or by the consumer (e.g., in a car trunk or in an insufficiently cold refrigerator).

Because the actual time to consumption for nitrite-cured products varies widely and residual nitrite decreases during storage, possibly to below the concentrations detectable with commonly used methods (<10 mg/kg), great difficulty is encountered in using residual nitrite content as an approach to regulation on this issue.

THE EFFECTS OF NITRITE IN CURED MEATS

The ability to promote the development of a reddish-pink color in cured meats was attributed to nitrate salts (saltpeter) by the late 1800s (Smith, 1895). It was later discovered that nitrite, and not nitrate, produced the typical color of cured meats (Kisskalt, 1899; Lehmann, 1899). The specific contribution of nitrite to the antimicrobial effects of cure salt was not recognized until the late 1920s (Lewis and Moran, 1928), and definitive evidence that nitrite was an effective antibotulinal agent came even later (Steinke and Foster, 1951). The role of nitrite as an inhibitor of lipid oxidation and its contribution to flavor have only recently been investigated (Cho and Bratzler, 1970; Sato and Hegarty, 1971).

---

[1]Subjecting of the product to a temperature more favorable to the growth of contaminating microorganisms than is the recommended storage temperature.

Inhibition of Microorganisms

The method used to preserve a particular product frequently determines the potential for the growth of spoilage and pathogenic microorganisms. Thus, the specific contribution of nitrite to the inhibition of potential pathogens and spoilage microorganisms varies with the products in which it is used and with variations in their production, handling, and abuse.

In its first report (National Academy of Sciences, 1981), the Committee reviewed data on the efficacy of nitrite as an antimicrobial agent and concluded that nitrite, in association with other components in the cure salt, exerts a concentration-dependent antimicrobial effect in cured products, including, but not limited to, inhibition of the outgrowth of spores of putrefactive and pathogenic clostridia, such as *Clostridium botulinum*. Thus, nitrite provides protection against the risk to health posed by botulism. Under conditions of excessive contamination or prolonged temperature abuse, nitrite does not indefinitely prevent such outgrowth, and spoilage or toxin production may ultimately ensue. Residual nitrite appears to be an important determinant of the degree of protection provided by nitrite. Thus, any product or process changes that result in a lower concentration of residual nitrite (e.g., adding less nitrite or increasing its rate of depletion) will increase the likelihood of toxicity if the product is contaminated and abused. However, other factors that influence the risk of botulism, such as contamination or the timing and duration of temperature abuse, are not predictable. It is therefore not possible to derive a quantitative relationship between the protection provided by nitrite and the risk of botulism or to determine the degree of protection that is necessary to ensure the safety of a particular product. However, a substantial amount of information exists in the case of some products, e.g., bacon, and for these, it is often possible to qualitatively predict the effect of changes in product formulation on the degree of protection against toxigenesis from *C. botulinum* spores.

A summary of the antimicrobial activity of nitrite in cured meats is given in Table 2-3. Depending on a number of factors--including the concentration of nitrite, environmental conditions, and the type of food product--nitrite may also contribute to the control of pathogens other than *C. botulinum*, e.g., *Staphylococcus aureus*, *Bacillus cereus*, and *C. perfringens*. However, nitrite is not regarded as a key factor in their control. In addition, nitrite retards microbial spoilage of cured meats by inhibiting the growth of a variety of organisms, especially anaerobic and aerobic spore-forming bacteria.

The mechanisms by which nitrite inhibits spore outgrowth and the growth of vegetative cells of microorganisms is not fully understood; but, in the case of some bacteria, it appears to involve reaction with iron-containing enzymes. A more thorough knowledge of the mechanism would facilitate a search for alternative antimicrobial agents.

TABLE 2-3

The Effects of Nitrite in U.S. Cured Red Meat and Poultry Products

Products[a]	Recommended Storage Temperature, °C[b]	Potential for Temperature Abuse or Contamination by Processor (P), Distributor (D), Consumer Before (CB) or After (CA) Opening	Spoilage Microorganisms: Reduced or Controlled by NO_2^-	Spoilage Microorganisms: Not, or Poorly, Controlled by NO_2^-	Inhibited by:[c]
Raw, Cured Products[f]					
High Water Activity:					
Bacon	<4.4	D - Low CB - Low CA - Very low	Aerobic mesophiles Corynebacteria[g]	Lactics[h] Micrococci	Low temperature NaCl Anaerobic packaging NO_2^- pH
Other pickle-cured products (e.g., smoked ham)	<4.4	D - Low CB - Moderate CA - High	Clostridia and Bacilli	Psychrotrophs[k] Lactics[h] Micrococci	Heating[f] followed by low temperature NO_2^- NaCl Anaerobic packaging pH
Low Water Activity:					
Dry-cured cuts (e.g., country ham)	0 to ambient (i.e., ∿15-35)	D - Moderate[l] CB - Moderate[l] CA - Moderate[l]	Clostridia	Molds Yeasts	Low a_w NO_2^- NaCl Anaerobic packaging
Dry, semidry, and fermented sausage (e.g., Lebanon bologna, salami)	0 to ambient (i.e., ∿15-35)	P - Moderate CB - Low CA - Low		Molds Yeasts	Acid or low a_w Heating NO_2^- Anaerobic packaging
Cooked, Cured Products					
Packaged After Heating:[m]					
Sausages (e.g., beef or chicken frankfurters) and some cold cuts	<4.4	P or D - Low CB - Low CA - Very low	Psychrotrophs[k]	Psychrotrophs[k] Lactics[h] Yeast	Heating followed by low temperature Anaerobic packaging
Canned:					
Perishable (e.g., canned ham)	<4.4	D - Moderate CB - High CA - High	Clostridia and other putrefactive anaerobes	Some clostridia	Pasteurization with NO_2^- followed by low temperature Sealed container NaCl
Shelf stable (e.g., luncheon meat)	Ambient (i.e., ∿15-35)	P or D - Very low CB - Extremely low CA - Low	Clostridia, thermophiles		Thermal process Sealed container NO_2^- NaCl
Commercially sterile (e.g., deviled ham)	Ambient (i.e., ∿15-35)	P or D - Extremely low CB - Negligible CA - Very low	Spore-formers (with faulty processing)		Thermal process Sealed container (NO_2^-/NaCl if processing faulty)

[a]Further examples of products of these categories can be found in the text.
[b]Most U.S. producers recommended <4.4° C (40° F), but <7.2° C (45° F) is still within some local laws.
[c]The rankings in Table 2-3 of the relative importance of the various factors inhibiting spoilage organisms or pathogens are judgments of the Committee based on data pertaining to U.S. products and, in a few cases, on data from studies in other countries.
[d]Color fixation of nitrite is selective for the muscle tissue of meat products.
[e]References to nitrite's contribution to flavor in various product classes are as follows: bacon: Huhtanen et al., 1981; Kimoto et al., 1976; MacDougall et al., 1975; Racquette et al., 1980; Wasserman et al., 1977; Williams and Greene, 1979; ham and ham-based products, including those canned: Brown et al., 1974; DuBose et al., 1981; MacDonald et al., 1980; dry-cured cuts: Eakes and Blumer, 1975; Eakes et al., 1975; Kemp et al., 1974; fermented sausages: Dethmers et al., 1975; frankfurters: Simon et al., 1973; Wasserman and Talley, 1972.
[f]Some raw cured products may be mildly heated during smoking (International Commission on Microbiological Specifications for Foods, 1980, pp. 136-159, 333-409; Nitrate Safety Council, 1980).

TABLE 2-3 (continued)

Pathogens: Reduced or Controlled by NO_2^-	Pathogens: Not, or Poorly, Controlled by NO_2^-	Pathogens: Inhibited by[c]	Effects of Nitrite on: Color[d]	Effects of Nitrite on: Flavor[e]	Effects of Nitrite on: Lipid Oxidation
C. botulinum Staphylococci[i]	Staphylococci[i]	Low temperature NO_2^- Fermentable carbohydrate (if added) NaCl Frying[j]	Color fixation	Inconsequential contribution (salt major contributor)	Inhibits
C. botulinum Staphylococci[i] *B. cereus*	Staphylococci[i] Salmonellae	Low temperature NO_2^- NaCl Anaerobic packaging[l]	Color fixation	Important contribution (salt also major contributor)	Inhibits
C. botulinum		Low a_w NO_2^- NaCl Anaerobic packaging[i]	Color fixation	Inconsequential contribution (salt and lipid oxidation major contributors)	May limit after some oxidation has occurred
C. botulinum Staphylococci[i]	Staphylococci[i] Salmonellae	Acid or low a_w	Color fixation	Important contribution (lactic acid production and salt also major contributors)	Inhibits
C. botulinum		Heating Low temperature NO_2^- Fermentable carbohydrate (if added) Packaging NaCl	Color fixation	Important contribution, if product not smoked; NO_2^- inconsequential, if smoked (spices also contribute)	Inhibits
C. botulinum		NO_2^- Low temperature Heating Sealed container NaCl	Color fixation	Important contribution[n]	Inhibits
C. botulinum		Thermal process with NO_2^- Sealed container	Color fixation	Important contribution[n]	Inhibits
C. botulinum		Thermal process Sealed container (NO_2^-/NaCl if processing faulty)	Color fixation	Important contribution[n]	Inhibits

[g]See National Academy of Sciences, 1981. Dempster (1980) studied corynebacteria sensitive to heating at 63°C for 30 min.
[h]Lactics include lactic-acid-producing bacteria, "mainly atypical streptococci" whose speciation remains to be satisfactorily defined (Reuter, 1975).
[i]See National Academy of Sciences, 1981; Gola and Casolari, 1979; Labots, 1976.
[j]Frying or other cooking by consumer or processor (as with prefried bacon).
[k]Psychrotrophs could include gram-negative bacteria, such as pseudomonads and coliforms, as well as yeasts (Terrell, 1974).
[l]This ranking is for packaged slices; for whole hams the potential is very low.
[m]Thus, subject to possible contamination during packaging.
[n]Many of these products are ham-based; thus, studies on ham flavor are pertinent (see footnote e).

Special considerations are relevant to the use of nitrite in fish products. Most important among these are the higher frequency of contamination of fish with *C. botulinum* and the fact that the most common contaminating strains are able to grow at lower temperatures (> 3.3).

Inhibition of Lipid Oxidation

Oxidation of lipids in meat changes its flavor and contributes to the development of rancidity during storage. Some products of lipid oxidation may also have adverse effects on human health.

Nitrite has been shown to retard lipid oxidation in cooked, noncured and cured meats. Nitrite may be especially important in comminuted products, where air incorporated during manufacture speeds the oxidative process. The mechanism of nitrite's inhibition of lipid oxidation is not thoroughly understood. Pearson *et al.* (1977) suggested that nitrite may stabilize the lipid components of the cell membranes in meat, or it may work by inhibiting the oxidative catalysts that are naturally present in meat.

The effects of nitrite on lipid oxidation in cured meats are summarized in Table 2-3 and are discussed in greater detail in Chapter 6.

Contribution to Sensory Characteristics of Cured Meats

Nitrite reacts with the myoglobin in meat to produce a reddish-pink color readily recognized as cured-meat color by consumers. The mechanism by which nitrite exerts this effect is discussed in Chapter 7.

The contribution of nitrite to flavor has not been fully determined for all cured products. Table 2-3 contains the Committee's judgment of its contribution to the flavor of several product classes.

Sodium chloride is largely responsible for the "cured" flavor in some products, especially bacon. Other product ingredients (e.g., spices) or processes (e.g., smoking) may also be important contributors to flavor. It is possible that flavor characteristics may not be attributable to a specific chemical component of cured meat, because the olfactory system can "synthesize," or "fuse," individual aromas into one that is different from that of the individual components.

The role of nitrite and other factors in the production of flavor in cured meats is discussed in more detail in Chapter 9.

Shortcomings of Nitrite

As currently used, nitrite shows little or no activity against some species, genera, or types of organisms that are potential health hazards or cause spoilage, e.g., lactobacilli, salmonellae, staphylococci, yeasts, and molds. The lack of activity against lactic-acid-producing bacteria, however, may not be totally disadvantageous, inasmuch as these organisms can metabolize fermentable carbohydrate to produce lactic acid. The lactic acid lowers the pH, and that results in inhibition of clostridial spore outgrowth and cell multiplication. This provides some protection against botulinal toxin production.

Because of interest in the fate of nitrite in cured meat, "bookkeeping" or "balance" experiments have been attempted to determine, at least generally, how much of the nitrite stays in the meat and how it is distributed. The studies that have addressed these problems are described more fully elsewhere (Cassens *et al.*, 1977; National Academy of Sciences, 1981).

Sebranek *et al.* (1973) concluded that the added nitrite was changed rapidly to other compounds and that little of it escaped from the product in volatile form. After processing, the change continued, but at a low rate, until the concentration of residual nitrite was low. Further work (Goutefongea *et al.*, 1971; Woolford and Cassens, 1977; Woolford *et al.*, 1976) indicated that the percentages of nitrite originally added and recovered in a given portion or fraction of meat shortly after processing are as follows: with myoglobin, 5-15%; as nitrite, 5-20%; as nitrate, 1-10%; as gas, 1-5%; with sulhydryl groups, 5-15%; with lipid, 1-5%; and with nonheme protein, 20-30% (Cassens *et al.*, 1977). It is not known whether these percentages change as the time after processing increases.

Because of its reactivity with meat components, only about 50% of the nitrite added to meat is detectable immediately after processing as residual nitrite by the usual analytic methods. There is uncertainty as to the chemical forms of residual nitrite and the forms and reactivity of the added nitrite that is not detectable. The residual nitrite undergoes depletion as the product is stored. The depletion is faster at higher temperatures--e.g., room temperature--than under refrigeration (Nordin, 1969). This depletion must be regarded as a shortcoming of nitrite. Because dormant spores remain viable for long periods (Pivnick *et al.*, 1969), it is desirable that the protective activity or potential be maintained, or at least deplete slowly, under normal storage and abuse conditions.

The characteristic color of cured red meats that nitrite "fixes" is susceptible to fading, if the product is exposed to oxygen (air) or light (Chapter 7).

EVALUATING ALTERNATIVES TO NITRITE

In approaching its task of evaluating alternatives to the use of nitrite, the Committee arrived at several general guidelines:

• Alternatives to nitrite must be usable in cured meats, poultry, and fish and amenable to production in the required amounts. Meat, in particular, is a complex medium and may contain a substantial proportion of lipid. The use of alternatives whose chemical properties differ radically from those of nitrite may require new techniques of application or processing. The capacity of alternatives to be used in cooked products to produce the desired effects, e.g., antimicrobial activity, should not be compromised by the heating these products receive.

• Ideally, an alternative or its capacity to produce its effects should deplete slowly or not at all during storage, because products may be consumed long after production.

• The Committee has avoided judgments on economic impacts and the consumers' desire for cured products, because those matters were not part of its charge. It sought alternatives to the current use of nitrite that would maintain the variety and types of products now available while ensuring adequate protection of the public health from microbial hazards.

• Because of the diversity of products in which nitrite is used and the variation in their handling and time to consumption, it is difficult to specify performance standards that should be met generally. Rather, alternatives are most easily judged by comparison with conventional nitrite use.

• As mentioned earlier in the report, the risks associated with the use of nitrite in cured meats could be reduced by decreasing the concentrations of nitrite added to these products, by using another method or agent to substitute for nitrite, or by incorporating inhibitors of the formation of *N*-nitroso compounds in products containing nitrite. The Committee has not attempted to quantitate the benefits of each means of risk reduction, e.g., reduced nitrite ingestion versus reduced intake of nitrosamines. However, in its first report it provided rough guides to the relative risks posed by various sources (National Academy of Sciences, 1981).

Nitrite fixes meat color, inhibits some microorganisms, inhibits lipid oxidation, and may contribute to the flavor of some cured meats. Most research on alternatives has been directed toward investigating agents that produce one of these effects--the activity most frequently tested is the inhibition of spore outgrowth and microbial proliferation, especially of *C. botulinum* (Table 2-4). Alternatives have also been suggested for the color-fixing and antioxidant activities of nitrite. However, to the Committee's knowledge, no substitutes have been suggested for nitrite's contribution to cured-meat flavor. In its assessment, the Committee evaluated each alternative first for its individual ability to exert the necessary antimicrobial effect, to produce the desired color, or to inhibit lipid oxidation independently. Combinations of agents (or treatments), such as irradiation

TABLE 2-4

Some Proposed Alternatives to Nitrite Evaluated by the Committee

Effect	Single Agents or Treatments	Combinations
Antimicrobial	Potassium sorbate or sorbic acid	Potassium sorbate or sorbic acid with low concentrations of nitrite
	Sodium hypophosphite	Sodium hypophosphite with low concentrations of nitrite
	Fumarate esters	
	Lactic-acid-producing organisms	Lactic-acid-producing organisms with low concentrations of nitrite
	Irradiation	Irradiation with low concentrations of nitrite
	Sodium chloride	
	Parabens	
	Glyceryl monolaurate	
	Humectants	
Lipid oxidation	Low concentrations of nitrite	
	Butylated hydroxytoluene	
	Citric acid	
	Ascorbic acid	
	α-Tocopherol	
Color production	Low concentrations of nitrite	
	Smoking	
	Nonabsorbable food colors	
	Nicotinates	
	Pyridine compounds	
	Betalines	
Inhibition of nitrosamine formation	Ascorbic acid	Ascorbic acid and α-tocopherol
	α-Tocopherol	

with low concentrations of nitrite, have also been tested for the ability to produce one or more of the effects attributed to nitrite. Where appropriate, the Committee evaluated such combinations (Table 2-4).

Alternatives identified by the Committee as the most promising were then evaluated for their sensory effects and toxicity. The results of these evaluations are given in Chapter 9 and 10, respectively. Long-term research strategies and methodological considerations are discussed in Chapter 11.

REFERENCES

American Meat Institute. 1980. Meatfacts. American Meat Institute, Washington, D.C. 27 pp.

Binkerd, E. F., and O. E. Kolari. 1975. The history and use of nitrate and nitrite in the curing of meat. Food Cosmet. Toxicol. 13:655-661.

Brown, C. L., H. B. Hedrick, and M. E. Bailey. 1974. Characteristics of cured ham as influenced by levels of sodium nitrite and sodium ascorbate. J. Food Sci. 39:977-979.

Cassens, R. G., G. Woolford, S. H. Lee, and R. Goutefonga. 1977. Fate of nitrite in meat. Pp. 95-100 in B. J. Tinbergen and B. Krol, eds. Proceedings of the Second International Symposium on Nitrite in Meat Products, PUDOC, Wageningen, the Netherlands.

Cerveny, J. G. 1980. Effects of changes in the production and marketing of cured meats on the risk of botulism. Food Technol. 34:240-243.

Cho, I. C., and L. J. Bratzler. 1970. Effect of sodium nitrtite on flavor of cured pork. J. Food Sci. 35:668-670.

Dempster, J. F. 1980. The microbiology of "back bacon" manufactured in "nitrate-free" brines after storage at 5°C. Pp. 272-275 in Proceedings of the 26th European Meeting of Meat Research Workers, Volume 2. American Meat Sciences Association, Chicago, Illinois.

Dethmers, A. E., H. Rock, T. Fazio, and R. W. Johnston. 1975. Effect of added sodium nitrite and sodium nitrate on sensory quality and nitrosamine formation in thuringer sausage. J. Food Sci. 40:491-495.

DuBose, C. N., A. V. Cardello, and O. Maller. 1981. Factors affecting the acceptability of low-nitrite smoked, cured ham. J. Food Sci. 46:461-463.

Eakes, B. D., and T. N. Blumer. 1975. Effect of various levels of potassium nitrate and sodium nitrite on color and flavor of cured loins and country-style hams. J. Food Sci. 40:977-980.

Eakes, B. D., T. N. Blumer, and R. J. Monroe. 1975. Effect of nitrate and nitrite on color and flavor of country-style hams. J. Food Sci. 40:973-976.

Gola, S., and A. Casolari. 1979. Antimicrobial activity of nitrite-dependent compounds. Pp. 267-279 in B. Jarvis, ed. Food Microbiology and Technology: Proceedings of the International Meeting on Food Microbiology and Technology, Medicina viva Servizio Congress, Srl., Parma, Italy.

Goutefongea, R., R. G. Cassens, and G. Woolford. 1977. Distribution of sodium nitrite in adipose tissue during curing. J. Food Sci. 42:1637-1641.

Gray, J. I., D. M. Irvine, and Y. Kakuda. 1979. Nitrates and N-nitrosamines in cheese. J. Food Prot. 42:263-272.

Health and Welfare Canada. 1975. Food and Drug Regulations, Amendment. Privy Council Order No. 1975-774. Canada Gazette Part II, 109:757-762 (April 23, 1975).

Huhtanen, C. N., F. B. Talley, J. Feinberg, and J. G. Phillips. 1981. Sensory and antibotulinal evaluation of sorbic acid containing bacon. J. Food Sci. 46:1796-1800.

International Commission on Microbiological Specifications for Foods. 1980. Microbial Ecology of Foods, Volumes 1 (pp. 1-332) and 2 (pp. 332-998). Academic Press, New York.

Kemp, J. D., J. D. Fox, and W. G. Moody. 1974. Cured ham properties as affected by nitrate and nitrite and fresh pork quality. J. Food Sci. 39:972-976.

Kimoto, W. I., A. E. Wasserman, and F. B. Talley. 1976. Sensory evaluation of flavor development in lean and adipose tissues of bacon. Lebensm.-Wiss. Technol. 9:274-276.

Kisskalt, K. 1899. Beitrage zur Kenntris der Ursachen des Rotwerdens de Fleisches beim Kochen nebst, einigen Versuchen uber die Wirkung der Schwefligen. Saure auf der Fleischforbe. Arch. Hyg. Bakt. 35:11.

Komarik, S. L., D. K. Tressler, and L. Long. 1974. Food Products Formulary Volume 1: Meats, Poultry, Fish, and Shellfish. AVI Publishing Company, Westport, Connecticut. 348 pp.

Kramlich, W. E., A. M. Pearson, and F. W. Tauber. 1973. Processed Meats. AVI Publishing Company, Westport, Connecticut.

Labots, H. 1976. Effect of nitrite on the development of _Staphylococcus aureus_ in fermented sausages. Pp. 21-27 in B. J. Tinbergen and B. Krol, eds. Proceedings of the Second International Symposium on Nitrite in Meat Products, Zeist, PUDOC, Wageningen, the Netherlands.

Lehmann, K. B. 1899. Uber das Haemorrhodin. Ein neues weit verbreitetes Blutfarbstoffderivat. Sber. Phys.-Med. Ges. Wurzb. 4:57.

Lewis, W. L., and J. A. Moran. 1928. The present status of our knowledge of ham souring. Pp. 1-141 in Bull. No. 4, American Institute of Meat Packers, Chicago, Illinois.

MacDonald, B., J. I. Gray, D. W. Stanley, and W. R. Usborne. 1980. Role of nitrite in cured meat flavor: Sensory analysis. J. Food Sci. 45:885-904.

MacDougall, D. B., D. S. Mottram, and D. N. Rhodes. 1975. Contribution of nitrite and nitrate to the colour and flavour of cured meats. J. Sci. Food Agric. 26:1743-1745.

National Academy of Sciences. 1981. The Health Effects of Nitrate, Nitrite, and N-Nitroso Compounds. National Academy Press, Washington, D.C. 544 pp.

National Oceanic and Atmospheric Administration. 1980. Processed Fishery Products, Annual Summary, 1979. Current Fisheries Statistics No. 8003. Department of Commerce, Washington, D.C.

Nitrite Safety Council. 1980. A survey of nitrosamines in sausages and dry-cured meat products. Food Technol. 34:45-51, 53, 103.

Nordin, H. R. 1969. The depletion of added sodium nitrite in ham. Can. Inst. Food Technol. J. 2:79-85.

Paquette, M. W., M. C. Robach, J. N. Sofos, and F. F. Busta. 1980. Effects of various concentrations of sodium nitrite and potassium sorbate on color and sensory qualities of commercially prepared bacon. J. Food Sci. 45:1293-1296.

Pearson, A. M., J. D. Love, and F. B. Shorland. 1977. "Warmed-over" flavor in meat, poultry, and fish. Adv. Food Res. 23: 1-74.

Pivnick, H., H. W. Barnett, H. R. Nordin, and L. J. Rubin. 1969. Factors affecting the safety of canned, cured, shelf-stable luncheon meat inoculated with Clostridium botulinum. Can. Inst. Food Technol. J. 2:141-148.

Price, J. F., and B. S. Schweigert. 1971. The Science of Meat and Meat Products, 2nd Ed. W. H. Freeman and Company, San Francisco, California.

Reuter, G. 1975. Classification problems, ecology and some biochemical activities of lactobacilli of meat products. Pp. 221-229 in J. G. Carr, C. V. Cutting, and G. C. Whiting, eds. Lactic Acid Bacteria in Beverages and Food. Academic Press, New York.

Sato, K., and G. R. Hegarty. 1971. Warmed-over flavor in cooked meats. J. Food Sci. 36:1098-1102.

Sebranek, J. G., R. G. Cassens, W. G. Hoekstra, W. G. Winder, E. Y. Podebradsky, and E. W. Kielsmeier. 1973. N-Tracer studies of nitrite added to a comminuted meat product. J. Food Sci. 38:1120-1126.

Simon, S., D. E. Ellis, B. D. MacDonald, D. G. Miller, R. C. Waldman, and D. O. Westerberg. 1973. Influence of nitrite and nitrate curing ingredients on quality of packaged frankfurters. J. Food Sci. 38:919-923.

Smith, E. 1895. Foods. D. Appleton and Company, New York.

Sofos, J. N., and F. F. Busta. 1980. Alternatives to the use of nitrite as an antibotulinal agent. Food Technol. 34:244-251.

Steinke, P. K. W., and E. M. Foster. 1951. Botulinum toxin formation in liver sausage. Food Res. 16:477-484.

Terrell, R. N. 1974. Theory and practice of preblending for use in comminuted meat food mixes. Pp. 23-33 in Proceedings of Meat Industry Research Conference. American Meat Institute Foundation, Washington, D.C.

U.S. Department of Agriculture. 1980. Statistical Summary: Federal Meat and Poultry Inspection for Calendar Year 1979. Food Safety and Quality Service, U.S. Department of Agriculture, Washington, D.C. 24 pp.

Wasserman, A. E., and F. Talley. 1972. The effect of sodium nitrite on the flavor of frankfurters. J. Food Sci. 37:536-538.

Wasserman, A. E., W. Kimoto, and J. G. Phillips. 1977. Consumer acceptance of nitrite-free bacon. J. Food Prot. 40:683-685.

Williams, J. C., and B. E. Greene. 1979. Plate waste of bacon cured with and without nitrite. J. Food Sci. 44:1260-1262.

Woolford, G., and R. G. Cassens. 1977. The fate of sodium nitrite in bacon. J. Food Sci. 42:586-589, 596.

Woolford, G., R. G. Cassens, M. L. Greaser, and J. G. Sebranek. 1976. The fate of nitrite: Reaction with protein. J. Food Sci. 41:585-588.

CHAPTER 3

PRESERVATION OF MEAT AND FISH PRODUCTS: GENERAL CONSIDERATIONS

CHAPTER 3

PRESERVATION OF MEAT AND FISH PRODUCTS: GENERAL CONSIDERATIONS

Because a comprehensive assessment of antimicrobial alternatives to nitrite requires evaluation of various preservation methods, such as irradiation and refrigeration, as well as the use of antimicrobial chemicals, this chapter identifies some of the general methods and objectives of food preservation pertinent to cured meats, poultry, and fish. Table 3-1 illustrates the mechanisms through which preservation techniques achieve their objectives in foods. Because the types of microorganisms that occur in foods are diverse, the table is generalized--not all microorganisms are controlled by each mechanism. Seemingly diverse methods of food preservation are based on a few factors that influence microbial proliferation. Some methods exert control through more than one mechanism.

Table 3-2 illustrates some of the control factors or mechanisms that operate in cured meats and some smoked fish. The symbols indicate that, for organisms that are affected, a particular control mechanism operates. They do not signify that all the microorganisms that might occur in that product are inhibited by a particular mechanism. Some factors singly often minimize microbial growth effectively. Usually, however, factors interact to reduce proliferation.

Nitrite is added to cured meats and fish as an antimicrobial agent to protect against pathogenic risk and to prevent product spoilage. Many intrinsic product factors influence the activity of nitrite or interact with it (National Academy of Sciences, 1981, Chapter 3; International Commission on Microbiological Specifications for Foods, 1980, pp. 136-159; Roberts _et al_., 1981a,b,c). The interactions among various factors may be additive, synergistic, or even antagonistic. The size or combination of barrier needed to minimize the growth of a potential pathogen or spoilage microorganism depends on the organism and its physiologic state (e.g., potentially injured by heating) and on the other factors present. In addition, the need for a preservative, such as nitrite varies among different products on the basis of other factors, such as the degree of contamination present and the likelihood of the product's being subjected to conditions favorable for microbial proliferation.

In this chapter, factors that influence microbial contamination or proliferation (and, hence, the need for nitrite) are reviewed for several categories of cured meats. This information is then used to

TABLE 3-1

Mechanisms Whereby Preservation Methods
Control Microbial Proliferation in Foods

Predominant Effect on Microbial Contaminants	Cause[a]	Examples of Preservation Methods[b]
Inactivation or injury	Increased temperature	Thermal processing, pasteurization
	Radiation	Radappertization (radiation)
Inhibited[a]	Low temperature	Refrigeration, freezing
	Low water activity	Drying, salting, curing, sugar addition
	Low pH	Acidulation, some fermentations
	Chemical inhibition	Curing, salting, natural smoking
	Microbial competition	Fermentations
	Gaseous atmosphere[c]	Vacuum packaging

[a]Some of the causes of inhibition, e.g., salting, may also cause death of some microbial species.

[b]Some methods act by more than one method or indirectly, e.g., fermentation can be viewed as microbial competition that lowers the pH or produces alcohol, which acts as a preservative.

[c]Unfavorable redox potential may also inhibit microbial proliferation but is not directly manipulated in any current preservation method.

TABLE 3-2

Influences on Microbial Survival or Proliferation for Various Cured Red Meat, Poultry, and Fish Products[a]

	Heating[b]		Inhibition of Undesirable Microbial Proliferation by					
	Inactivation	Injury	Low Temperature[c]	a_w	pH	Preservatives[d]	Microbial Competition[e]	Vacuum Packaging[f]
Raw, cured meat products								
Bacon	+	0	++	+	0	+	+	+
Other pickle-cured products, e.g., hams	+	0	++	0	0	+	+	0
Dry-cured cuts, e.g., country hams	0	0	+	++	0	+	0	+
Dry, semidry, and fermented sausages	+	0	+	++	++	+	++	+
Cooked, cured meat products								
Perishable (pasteurized), packaged after heating, e.g., frankfurters	+	0	++	0	0	+	-	+
Perishable (pasteurized), canned, e.g., hams	+	0	++	0	0	+	-	+
Shelf-stable, canned, e.g., luncheon meats	+	+	0	0	0	++	-	0
Commercially sterile, e.g., deviled ham	++	+	0	0	0	0	-	0
Smoked Fish	+	0	++	0	0	+	-	+

[a]Key: - = Reduced effectiveness in controlling undesirable microbial proliferation.
0 = Not involved in controlling microbial proliferation.
\+ = Supplementary control on microbial proliferation.
++ = Major control on microbial proliferation.

[b]Some raw products are mildly heated. Vegetative cells of most bacterial species are inactivated by the temperatures used in heating bacon and other pickle-cured products, some fermented sausages, and cooked products. Inactivation of spores requires more severe heating as used for commercially sterile and shelf-stable products. Commercial sterilization results in a minimum of 12 log cycles reduction in the number of viable *C. botulinum* spores in products receiving a "botulinum" cook, but thermal processing of shelf-stable products yields only approximately a 3-log reduction. Remaining viable spores are damaged and, thus, are more sensitive to the effects of other controls.

[c]Refrigeration, if recommended.

[d]Sodium chloride and sodium nitrite. Preservatives in commercially sterile products may inhibit microbial proliferation if inadequate thermal processing has allowed viable cells or spores to survive.

[e]Naturally occurring (or in fermented sausages, added) lactic-acid-producing bacteria, which are generally inactivated by cooking, may lower pH, if fermentable carbohydrate is present.

[f]Vacuum packaging of cured products controls aerobic spoilage organisms, but generally does not affect proliferation of *C. botulinum*.

determine which products are in most need of the antimicrobial properties of nitrite. The information on product variation, especially with respect to intrinsic susceptibility to microbial proliferation, is also used by the Committee as an indication of which intrinsic factors may be manipulated to reduce the need for nitrite in various products.

PRODUCT CHARACTERISTICS INFLUENCING HEALTH RISKS

The frequency of disease outbreaks from various pathogens in meat and poultry products in recent years is shown in Table 3-3. Botulism is generally the most feared of foodborne diseases, undoubtedly because of its traditionally high fatality rate (Center for Disease Control, 1979). The rate has been decreasing in recent years (Morris and Hatheway, 1980). Information on the incidence of botulism and its causes was reviewed by the Committee in its first report (National Academy of Sciences, 1981, Chapters 2 and 10). Of the 15 outbreaks (39 cases, 16 deaths) of botulism attributed to commercially processed meat or poultry that occurred in the United States or Canada during the period 1899-1980, seven outbreaks (21 cases, six deaths) involved products that are normally cured (Tompkin, 1980).

The origins of the pathogenic organisms that occur in cured meats are listed in Table 3-4, with information on the factors that may control their growth.

The major health benefit of the use of nitrite is its inhibitory action on *Clostridium botulinum*, i.e., the protection it provides against botulism. The sequence of events leading to botulism is illustrated in Figure 3-1. The potential for botulism is determined largely by the following:

- Survival of spores in the finished product, because of intentionally mild heating or faulty thermal processing that results in suppression of competitive organisms.

- The capacity of the intrinsic characteristics of the finished product to control *C. botulinum* spore outgrowth and cell multiplication.

- Exposure of the product to temperatures that permit growth, i.e., ambient storage temperature for shelf-stable products and commercially sterile products that may have been inadequately thermally processed, or temperature abuse of perishable products.

In its first report, the Committee considered at length the possibility of expressing the protection provided by nitrite in terms of the number of cases of botulism avoided, i.e., attempting to relate the risk of botulism to the nitrite-induced delay in toxin production in temperature-abused cured products. However, it found that other factors that influenced the risk of botulism, such as

TABLE 3-3

Number of Meat and Poultry Products that were Reported as Foodborne Vehicles in Disease Outbreaks in the United States, 1968-1977[a]

Disease	Beef						Pork				Other meats		Meat, general				Poultry					Totals	
	Beef (specified)	Beef (unspecified, possible roast beef)	Roast beef	Ground beef	Cured beef	Foods containing beef	Pork	Cured pork (ham)	Sausage	Foods containing pork	Lamb	Other meats	Unspecified meats	Luncheon meats	Liver	Gravy	Turkey	Chicken	Duck	Poultry salad	Food containing poultry	Total outbreaks in which meat or poultry were vehicles	Percent of outbreaks in which meat or poultry were proven or suspicious vehicles
Arizonosis																	1					1	25
Bacillus cereus gastroenteritis		1				1				1			1								1	5	39
Botulism				1		5			1			6						2			1	16	14
Clostridium perfringens gastroenteritis	9	11	36	14	1	11	4		1		1		4	1		10	29	6			1	139	90
Salmonellosis	4	13	22	8	2	3	13	6	4			2	5	3	1	1	39	15	1	11	4	157	44
Shigellosis															1		1			2		4	8
Staphylococcal intoxication	9	4	7	7	4	4	23	105	7	5	1	2	5	6	2	3	23	16		16	3	252	8
Streptococcal group D gastroenteritis			3	1		1		2	1				1		1		1					11	65
Hepatitis A								1						2								3	9
Toxoplasmosis				1																		1	100
Trichinosis				7			39		42	6		6										100	94
Chemical poisonings	1			3			2			4												10	9
Diseases of unknown etiology	24	44	85	81	11	49	18	77	31	16	5	3	16	22	3	18	91	52	3	22	22	693	31
Total	47	73	153	123	18	74	99	191	87	32	7	19	32	34	8	32	185	91	4	51	32	1392	
		226																					
	488						409				26		74				363						
Percent	3.4	5.2	11.0	8.8	1.3	5.3	7.1	13.7	6.3	2.3	0.5	1.4	2.3	2.4	0.6	2.3	13.3	6.5	0.3	3.7	2.3		
		16.2																					
	35.1						29.4				1.9		5.3			4.9	26.1						

[a] Reprinted with permission from Bryan, 1980.

TABLE 3-4

Characteristics of Potential Pathogens Occurring in Cured Meats[a]

Organism	Reservoir	Vehicle of Transmission	Transmission	Disease Agent	Time, Temperature Requirements for Production of Toxic or Infective Levels	Growth Requirements
Clostridium botulinum Group I, proteolytic types A and B	Soil	Soil-derived dust, nonmeat ingredients	As spores	Neurotoxin	Days at 10-50°C (optimum, 37°C)	Anaerobic
Clostridium botulinum Group II, nonproteolytic types A, B, and F	Soil, sediments, water	Dust, water	As spores	Neurotoxin	Days at 3.3-50°C (optimum, 30°C)	Anaerobic
Staphylococcus aureus	Livestock, man	Nose, skin, lesions	As vegetative cells	Enterotoxin	5 h at 7.8-43°C (optimum, 37°C)	Aerobic, facultatatively anaerobic
Salmonellae	Animals, including man	Contact with feces	As vegetative cells	Vegetative cells, multiplying in gastrointestinal tract	Zero to several hours at 6.7-45.6°C (optimum, 37-43°C)	Aerobic, facultatively anaerobic

[a] *Clostridium perfringens* may be present, but is not a major cause of disease outbreaks involving cured meats.
[b] These values may differ if other growth conditions are suboptimal. Values are for growth in meat products, where data were available.
[c] Value is typical for cured meats, which contain salt.
[d] NA = not applicable.

TABLE 3-4 (continued)

Heat Resistance		Limiting Conditions[b] for Proliferation or Toxin Production				Effect of	
Organism	Toxin	Temperature, °C	Water Activity	Brine, %	pH	Microbial Competition	Sodium Nitrite (120-200 mg/kg)
High (spores)	Low	<10	<0.94	>8-10	5.0-5.6[c]	Very inhibitory, especially at low pH	Effective
Moderate (spores)	Low	<3.3	<0.95	>5	5.0-5.6[c]	Very inhibitory, especially at low pH	Effective
Low	High	<7.8	<0.87	>12-13	4.7	Very inhibitory, especially at low pH	May act anaerobically, not effective aerobically
Low	NA[d]	<6.7	<0.95	∿5.3	4.5	Inhibitory, especially lactic acid bacteria	Not effective

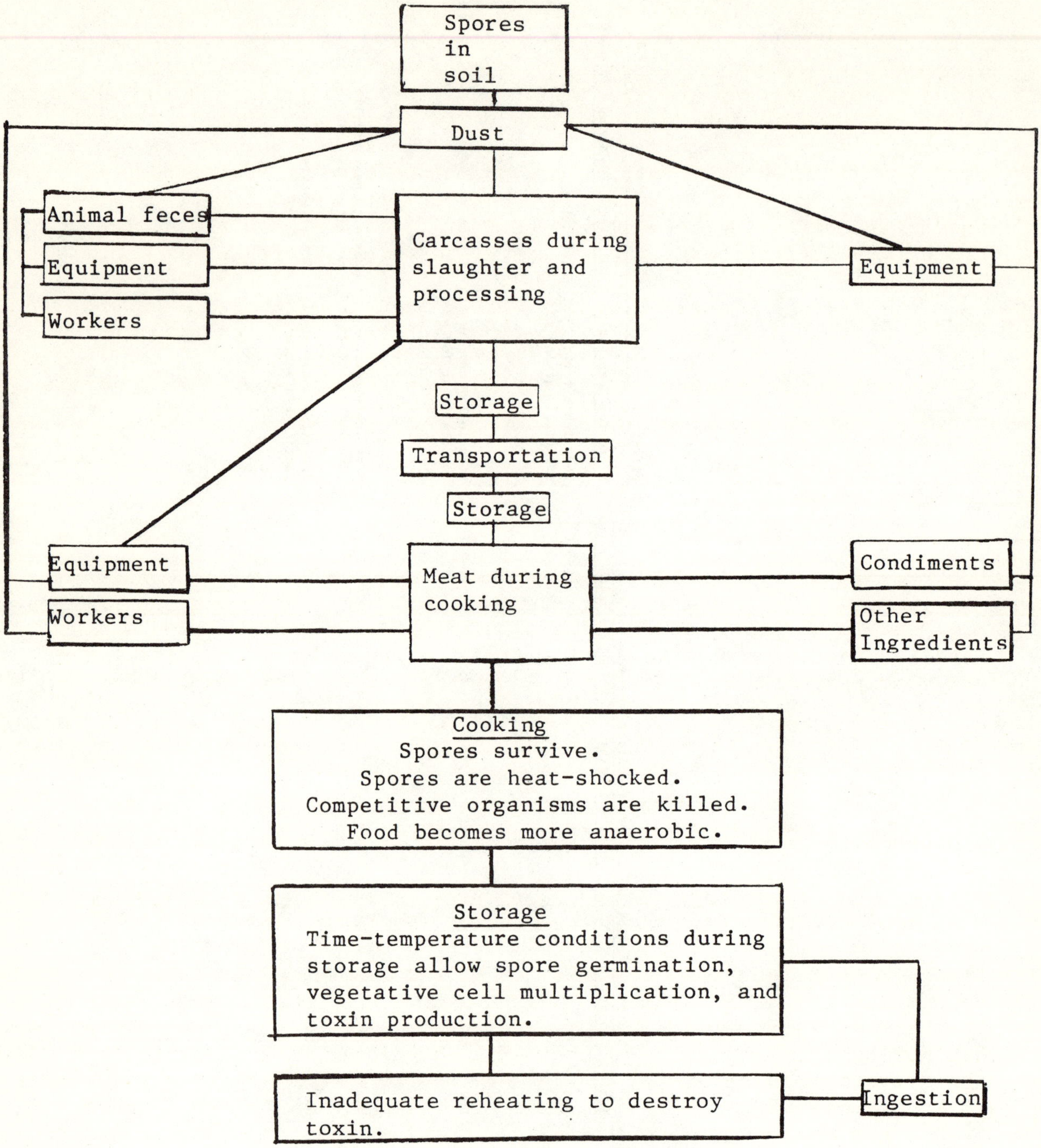

FIGURE 3-1. Web of causation of Clostridium botulinum food-borne intoxication. Modified from Bryan, 1979. For fish, the reservoir of spores is water or sediments.

product contamination and the timing and duration of temperature abuse, were not predictable (National Academy of Sciences, 1981, Chapter 10). Thus, it is not possible to derive a quantitative relationship between the protection provided by nitrite and the risk of botulism or to determine the degree of protection that is necessary to ensure the safety of a particular product. However, a substantial amount of information exists in the case of some products, e.g., bacon, and for these it is often possible to qualitatively predict the effect of changes in product formulation on the degree of protection against toxigenesis from _C. botulinum_ spores.

In addition, because products differ in the degree and frequency of microbial contamination, in the opportunities for microbial contamination or multiplication during production and handling, and in the extent to which control is exerted over pathogens by the intrinsic characteristics of the product other than nitrite (e.g., pH, brine concentration), the Committee believes that it is possible to rank product types according to the degree to which the addition of nitrite (or an alternative) is needed to reduce the risk of botulism.

Degree and Frequency of Product Contamination

Meat may be contaminated during slaughter, processing, and/or handling procedures. Equipment, personnel, additives, and other environmental contacts serve as possible reservoirs of contamination. In addition, microorganisms from the intestinal tract, the lymphatic system, and the outer surfaces of the animals may also contaminate meat products (Johnston and Elliot, 1976). After processing, distribution to retail outlets, and sale to the consumer, microbial contamination may occur during storage, food preparation, and handling immediately before consumption. The sources of contamination frequently determine the range of microbial contaminants present and, thus, the potential for pathogenic hazard or product spoilage. Initial contamination and the severity of thermal processing combine to determine the spore contamination of the finished product.

The contamination of different products with _C. botulinum_ spores has not been investigated extensively. Available data indicate that the average occurrence of _C. botulinum_ spores is about 1-2 spores per 10 kg of meat (Holley, 1978). Data collected in the United Kingdom by Roberts and Smart (1976) in serial samples from one producer indicate that "pockets" of contamination considerably higher than their average of 1-2 spores per kilogram of meat occurred occasionally. Their average is higher than the estimate of 0.064 spore/kg made by Hauschild and Hilsheimer (1980) for bacon samples sold at retail in Canada.

Abrahamsson and Riemann (1971) found that occurrence of spores in products sold at retail in the United States was highest in cooked hams (1.6 spores/kg), compared with turkey (0.8 spore/kg), smoked chicken (no spores in 1.5 kg), and other semipreserved meats (no spores in 5 kg). Surveys of fish indicate a higher degree of contamination (Sakaguchi, 1979, pp. 405-406). Because spores of *C. botulinum* are ubiquitous (Smith, 1977), the Committee believes that it is not practicable to guarantee that raw meat or fish will never contain spores.

Production, Distribution, Storage, and Handling

During the production of different cured products, various opportunities may be presented for increased contamination or for the multiplication of pathogens. Other opportunities for pathogen proliferation may occur during distribution or while the consumer is in possession of the product. Information on these possibilities is summarized in Table 3-5, which draws on information presented by Bryan (1980) on disease outbreaks associated with meat and poultry. Mishandling occurs predominantly in food-service establishments or in consumer possession (Bryan, 1980).

Assessment of the need for an alternative to the current use of nitrite in protecting health should take into account the various factors that contribute to the hazard and the extent to which they can be modified or eliminated by other means. For example, the staphylococcal hazard in cured products can be reduced considerably by adherence to good manufacturing practices (National Academy of Sciences, 1975; U.S. Department of Agriculture, 1977). The hazard from salmonellae can also be reduced considerably during processing and by good hygienic practices by the consumer (National Academy of Sciences, 1969). The International Commission on Microbiological Specifications for Foods (1980) and Genigeorgis and Riemann (1979) have reviewed practices that can be used to reduce hazards from foods that arise during production and distribution.

The surveillance and recall system for products potentially posing foodborne hazards probably further reduces risk to the consumer (Johnston and Krumm, 1980). However, this system is expensive.

Intrinsic Product Characteristics

A critical factor in evaluating the risk from cured products and the need for an alternative to nitrite is the extent to which the intrinsic characteristics of the products, such as water activity and pH, act to control the proliferation of pathogens. Table 3-6 summarizes the characteristics of U.S. cured products that may influence microbial growth. Comparison of Tables 3-6 and 3-4 indicates that an individual controlling factor in a cured product would rarely exclude the possibility of pathogen growth. Additionally, the effects of some factors that influence proliferation (e.g., meat type and cut, curing adjuncts) are poorly understood. Limitation of proliferation stems from the interactions of the various factors that control growth and is best judged from experimental evidence, because these may be additive, synergistic, or antagonistic.

TABLE 3-5

Influence of Production, Storage, Distribution, and Handling on Pathogenic Risk[a]

	Opportunities for Increased Contamination of Product Presented During		Opportunities Presented for Microbial Growth During				
	Processing	Packaging	Production	Recommended Storage	Distribution and Retailing	Consumer Handling	Food Preparation[b]
Raw, cured products							
Bacon	NI[c]	When handled for slicing	NI	NI	Infrequent	Infrequent	Usually fried or cooked
Other pickle-cured products, e.g., hams	NI	NI	NI	NI	Infrequent	Relatively frequent[d]	Not always cooked
Dry-cured cuts, e.g., country hams	NI	When handled for slicing	If manufacturing practices suboptimal	NI	Infrequent	Relatively frequent after cooking[d]	Not always cooked
Dry, semidry, and fermented sausages	From equipment surfaces during comminution	NI	If manufacturing practices suboptimal[d]	NI	Relatively frequent (but product characteristics control growth well)	Relatively frequent (but product characteristics control growth well)	Generally not cooked
Cooked, cured products							
Perishable (pasteurized), packaged after heating, e.g., frankfurters	From equipment surfaces during comminution	When handled for slicing after pasteurization	NI	NI	Infrequent	Varies with product--infrequent for frankfurters, relatively frequent for "cold cuts"	Varies with product--many types vary rarely cooked
Perishable (pasteurized), canned, e.g., hams	NI	NI	NI	NI	Relatively frequent	Relatively frequent[d]	Sometimes cooked
Shelf-stable, canned, e.g., luncheon meats	From equipment surfaces during comminution	NI	NI	If other controls --e.g., thermal processing --fail	If other controls --e.g., thermal processing --fail	Moderately frequent	Rarely cooked
Commercially sterile, e.g., deviled ham	From equipment surfaces during comminution	NI	NI	If other controls --e.g., thermal processing --fail	If other controls --e.g., thermal processing--fail	Moderately frequent	Very rarely cooked

[a]This table focuses on the risk from *Clostridium botulinum*, but similar (although not identical) considerations apply to *Staphylococcus aureus* and salmonellae.

[b]Sufficient cooking may inactivate botulinum toxin, but not staphylococcal enterotoxin. Some heating, e.g., during use of bacon as a flavoring agent, may not be adequate to destroy botulinum toxin. Many products are only very rarely, if ever, cooked.

[c]NI = no notable influence.

[d]Growth of staphylococci is the major problem under these circumstances.

TABLE 3-6

Intrinsic Characteristics of Cured Products Other Than Preservatives[a] that May Minimize the Proliferation of Pathogens[b,c]

	Water Activity	Brine, %	pH	Microbial Competition[d]
Raw, cured products				
Bacon	>0.95	3.0-6.0	∿(5.6-6.4)	Possible
Other pickle-cured products, e.g., hams	>0.95	3.0-6.0	∿(5.6-6.4)	Possible
Dry-cured cuts, e.g., country hams	∿0.92	Increases during production to ∿10 (internal)	∿(5.6-6.4)	Possible[f]
Dry, semidry, and fermented sausages:				
Cooked	∿0.95	∿(10-16)	4.8-5.2	Possible[e]
Semidried	<0.92	∿(10-16)	4.8-5.2	Encouraged[f]
Dried	<0.92	∿(10-16)	4.8-5.2	Encouraged[f]
Cooked, cured products				
Perishable (pasteurized), packaged after heating, e.g., frankfurters	>0.95	3.5-5.5	∿(5.6-6.4)	Possible
Perishable (pasteurized), canned, e.g., hams	>0.95	3.5-5.5	∿(5.6-6.4)	Unlikely (minimized)
Shelf-stable, canned, e.g., canned luncheon meats	>0.95	3.5-5.5	∿(5.6-6.4)	Unlikely (minimized)
Commercially sterile, e.g., deviled hams	>0.95	3.5-5.5	∿(5.6-6.4)	Unlikely (minimized)

[a]Sodium chloride and sodium nitrite.
[b]Potential pathogens of concern are *Clostridium botulinum*, *Staphylococcus aureus* and salmonellae.
[c]A number of other intrinsic factors undoubtedly affect microbial proliferation, e.g., meat species, type and cut, and curing adjuncts (such as phosphates and ascorbate), but much less is known about their influence.
[d]From natural flora if fermentable carbohydrate added.
[e]By back-inoculation or with starter culture.
[f]Because recontamination may occur during packaging.

Various methods have been developed for expressing the control exerted by a product's intrinsic characteristics and the contribution of nitrite to this control. One is to compare the time necessary to elicit toxin production in various nitrite-containing and nitrite-free products inoculated with a known number of *C. botulinum* spores and deliberately held at temperatures that permit growth. With data from experiments of this type, the probability of outgrowth of and toxigenesis from *C. botulinum* spores can be calculated for products that contain different concentrations of nitrite or other preservatives, if they are subjected to a specified period of temperature abuse. Comparison of such probabilities for nitrite-free products facilitates identification of products that are most susceptible to *C. botulinum* outgrowth and toxigenesis: the higher the probability, the greater the need for nitrite addition or some other method to reduce the likelihood of toxin formation.

Appendix D contains calculations of the probability of spore outgrowth and toxigenesis for various types of products. The probabilities for products free of added nitrite are summarized in Table 3-7. There are clearly considerable differences among product classes with respect to the extent to which a nitrite-free product would support toxigenesis if temperature-abused; but substantial differences can also occur within a product class (e.g., bacon). The Committee therefore believes that generalizations regarding the relative susceptibility of product classes to toxigenesis must always be made cautiously. Judgments should always take into account the diversity of formulations (e.g., with or without added sugar or ascorbate) and processing methods used with particular types of products.

Review of Tables D-1, D-2, and D-3 in Appendix D and Table 3-7 reveals that the probability of toxigenesis varies with the following:

- Type of product, e.g., fermented sausage versus turkey ham.
- Type of meat in a product, e.g., chicken versus beef or pork in frankfurters.
- Components of the cure other than nitrite, e.g., fermentable carbohydrate in bacon, salt concentration in liver sausages, and ascorbate or isoascorbate in canned perishable comminuted meats.

The relative susceptibilities of various classes of products to toxigenesis revealed that pasteurized products, particularly those with low salt concentrations and those made from poultry, are at greatest risk of toxigenesis under conditions of temperature abuse. Other products apparently have greater intrinsic resistance to toxigenesis, e.g., fermented sausage and some types of bacon have other factors (such as high salt concentration) that may limit pathogenic proliferation. However, unless fermentable carbohydrates and starter cultures of bacteria that convert the carbohydrates to acid are added to dry, and especially semidry, fermented sausages, nitrite (and/or nitrate) may be essential for safety.

TABLE 3-7

The Probability of Outgrowth and Toxigenesis or Swelling from *C. botulinum* Spores in Different Nitrite-Free Cured Meats[a]

Product Class	Product	Probability[b]
Raw, cured		
Bacon	Bacons	1×10^{-2}–$<6 \times 10^{-8}$
Fermented, dried and semidried sausages	Fermented, e.g., summer sausage	$<10^{-5}$
Cooked, cured		
Perishable (pasteurized), packaged after heating	Liver sausage (ms)[c]	$\geq 10^{-2}$–10^{-4}
	Turkey ham	$\geq 10^{-2}$–10^{-4}
	Chicken frankfurters	$\geq 10^{-2}$–10^{-4}
	Frankfurters	10^{-4}–$<10^{-5}$[d]
	Liver sausage (hs)[c]	$\leq 10^{-5}$
Perishable (pasteurized), packaged after heating	Comminuted pork	$>3 \times 10^{-3}$[e]
		$>3 \times 10^{-3}$–10^{-6}[f]
Commercially sterile		$<10^{-12}$

[a]Tests conducted under conditions of simulated temperature abuse for 1 week at 27°C.

[b]Probability, $p = (\ln n/q)/s$, where n = no. packages, q = no. nontoxic packages, and s = no. challenge spores/package. Data in this table are extracted from Tables D-1, D-2, and D-3 in Appendix D.

[c]ms = moderately salted, hs = highly salted.

[d]Includes some tests conducted with sodium nitrite at 30 mg/kg.

[e]Without added ascorbate or isoascorbate.

[f]With sodium ascorbate or isoascorbate at ∿200 mg/kg.

PRODUCT SUSCEPTIBILITY TO SPOILAGE

Table 3-8 illustrates that the spoilage problems in cured meats are characteristic of the type of product and are best considered for each product or type individually.

A wider range of microorganisms are involved in spoilage than in health hazard, and the organisms of concern differ more among products. In addition, less is known about the specific organisms involved, the factors that limit their growth, and the relative contributions of the various factors that control microbial growth.

Many of the considerations involved in protecting the aesthetic appeal of cured products against deleterious microbial growth are similar to those involved in preserving its freedom from microbial hazard to health. Spoilage is dealt with separately here, to emphasize the dual aspects of preservation and to point out that alternatives to nitrite aimed at protecting health may not match with alternatives for inhibiting spoilage.

The possibility of product spoilage is governed by considerations similar to those governing the possibility of pathogenic hazard:

- Degree of initial contamination of the product and the effects of processes used in production, e.g., thermal processing and temperatures that allow microbial growth.
- Capacity of the product itself to control growth of spoilage microorganisms.
- Exposure of the product to temperatures that permit microbial growth.

The contribution of nitrite to the control of spoilage in various categories of cured-meat products was discussed in the Committee's first report. The major antispoilage effect of nitrite is mediated through its inhibitory action on putrefactive anaerobic and aerobic spore-forming bacteria, such as clostridia and bacilli, which become a problem only when products are exposed to temperatures that permit their growth. Spores of putrefactive anaerobes outnumber those of *C. botulinum* in raw meats and cured products (Holley, 1978). They may be more resistant to the lethal effects of heat and other agents (International Commission on Microbiological Specificatons for Foods, 1980, pp. 1-37).

On the basis of considerations listed above, the products most likely to become spoiled if nitrite were omitted are shelf-stable and perishable canned products.

- Shelf-stable products do not receive thermal processing adequate to kill the spores of putrefactive spore-forming

TABLE 3-8

Microbial Defects of Red Meats and Their Products[a]

Product	Defect	Organism
Fresh meat		
Fresh, refrigerated (0-5°C)	Off-odor, slime, discoloration	Pseudomonas, Aeromonas, Alcaligenes, Acinetobacter, Microbacterium, Moraxella, Proteus, Flavobacterium, Alteromonas, Saccharomyces
	Lipolysis, pungent odor	Pseudomonas, yeasts
	Moldy	Penicillium
	Whiskery	Thamnidium
	Black spot	Cladosporium
	White spot	Sporotrichum
Fresh (15-30°C)	Bone taint	Clostridium
	Gassy	_C. perfringens_
	Foul odor	_C. bifermentans_, _C. histolyticum_, _C. sporogenes_
Vacuum-packaged	Acid, sweet, rancid	Lactobacillus, Microbacterium, Enterobacter, Hafnia
Cured meat		
Bacon (vacuum-packaged)	Cheesy, sour, rancid	Micrococcus
	Discoloration	Molds
	Slight souring	Lactobacillus, Micrococcus, Vibrio, Alcaligenes, Corynebacterium
	Putrefaction	_Clostridium sporogenes_
Other vacuum-packaged meats	Cabbage odor	_Proteus inconstans_
	Tainted	Vibrio
Brines	Turbid	Debaryomyces, Kloeckera
Ham	Surface slime	Micrococcus, Microbacterium, yeasts
	Gassy or puffy	Clostridium
	Green discoloration	Lactobacillus, Streptococcus, Leuconostoc
	Bone and meat "sours"	Clostridium
	Surface growth (dry-cured)	Molds

TABLE 3-8 (continued)

Product	Defect	Organism
Cured meat (continued)		
Sausages	Surface slime	Micrococcus, yeasts
	Gas production (vacuum-packaged)	
	Greenish discoloration	Lactobacillus viridescens, Leuconostoc
Fermented sausage	Slime	Yeasts
	Spots	Molds
Canned meat		
Commercially sterile	Gas, putrefaction	Spore-formers (Bacillus, Clostridium)
Semipreserved	Souring, discoloration	Streptococcus
	Putrefaction, gas	Bacillus, Clostridium

[a]Taken from Banwart, G. J. 1979. Basic Food Microbiology. AVI Publishing Company, Inc., Westport, Connecticut. 781 pp. With permission.

bacteria, and these products are stored at temperatures that permit spore outgrowth. Spoilage would occur if the thermal processing were not reinforced by the presence of sodium chloride and sodium nitrite.

- Omitting or reducing nitrite in perishable canned products, such as hams, would increase spoilage problems, because these products are more often subjected to temperature abuse, such as unrefrigerated storage, at retail or by the consumer.

If careful attention is paid to hygiene and equipment sanitation during production and packaging of products and to continuous and correct refrigeration, spoilage will not be a major problem. Reducing nitrite or omitting it from cured products may permit the growth, under refrigeration, of some psychrotrophs that would otherwise be inhibited (Terrell, 1974).

The interrelationship of spoilage and health hazard should be borne in mind during consideration of actions to control either. some spoilage microorganisms, such as lactobacilli, may exert an inhibitory effect on pathogen growth through microbial competition or other means, such as acid production from fermentable carbohydrate added to the product. However, products may become toxic before they are spoiled. Thus, spoilage cannot be relied on to decrease risks from pathogens by rendering consumption unlikely.

CONCLUSIONS

On the basis of the evaluation done by the Committee for its first report (National Academy of Sciences, 1981) and the data presented in this chapter, the Committee has considered which products are most likely to become toxic if nitrite, or an antimicrobial alternative, were not added. To develop an accurate ranking, one must know which products are most likely to contain spore contamination, which are most susceptible to toxigenesis, and which are most likely to be exposed to temperatures that favor toxigenesis. It is not possible to make judgments on the first of these (but suggestions for further studies are presented later); however, by combining evidence on the intrinsic susceptibility to toxigenesis of various types of cured products (Table 3-7) and on other influences on risk, such as frequency of exposure to temperatures permitting growth of pathogens (Table 3-5), the Committee has attempted to assess the relative risk from changes in nitrite addition to various product types.

The major product categories most likely to pose increased health hazards if nitrite were not added or were added at lower concentrations are canned shelf-stable cured products (given less than a botulinum cook), perishable canned and other hams (excluding dry-cured hams), and pasteurized poultry products. In addition, shelf-stable and perishable canned products are more likely to become spoiled if nitrite were omitted. Thus, these products are most in need of an alternative means of controlling microbial proliferation. Factors other than nitrite contribute considerably to the safety of dry and semidry fermented sausage, dry-cured cuts, and some bacons, especially those to which fermentable carbohydrate is added, and any possible increase in health hazard posed by reducing nitrite addition to these products would be less. However, unless fermentable carbohydrates and starter cultures are added, nitrite may be essential for the safety of fermented sausages.

Some approaches to the reduction of microbial proliferation or contamination, such as refrigeration or increased hygiene, would have results similar to those of nitrite addition, if they were rigorously applied. But they involve considerable uncertainties that arise from the possibility of human error or mechanical failure. Thus, the Committee believes that nitrite addition is most closely approximated by alternatives that function by killing contaminants in products--e.g., thermal processing or irradiation--or that are intrinsic to the products--e.g., preservative compounds, a_w, and pH.

RECOMMENDATIONS

1. If intrinsic factors in cured meats that exert control over microbial proliferation were optimized, the need for preservatives, such as nitrite or the alternatives considered in Chapter 4, might be reduced. To determine the extent to which this might be possible, the Committee reiterates a recommendation made in its first report:

> High priority should be accorded to investigations of the interaction of factors controlling pathogens and spoilage organisms in different commercial products in order to develop methods for predicting the degree of control gained or lost through alteration of any of those factors.

Variables to be investigated in such work include:

- Product composition, including cut of meat used and the species from which it was taken.
- Product characteristics, e.g., pH, a_w, E_h, and brine concentration.
- Types and concentrations of curing adjuncts, e.g., phosphates, ascorbate, antioxidants, chelators, and nitrosation inhibitors.
- Processing conditions, such as comminution (which may affect product uniformity) and temperature and duration of heat treatment.

- Atmosphere in which a product is enclosed.
- Incubation (simulated abuse) temperature.
- Storage (at various temperatures for different periods) before temperature abuse.
- Degree of contamination, e.g., spore load.

2. As a contribution to assessing the need for nitrite or an alternative, studies should be conducted to determine the extent to which *C. botulinum* spores contaminate different types of raw meats and products made under diverse conditions in different geographic locations.

3. The Committee recommends periodic evaluation of the need for nitrite or some alternative in various products or product types as new information (e.g., on spore contamination) becomes available or as processing procedures change.

REFERENCES

Abrahamsson, K., and H. Riemann. 1971. Prevalence of Clostridium botulinum in semi-preserved meat products. Appl. Microbiol. 21:543-544.

Banwart, G. J. 1979. Basic Food Microbiology. AVI Publishing Company, Westport, Connecticut. 781 pp.

Bryan, F. L. 1979. Epidemiology of food-borne diseases. Pp. 3-69 in H. Riemann and F. L. Bryan, eds. Food-borne Infections and Intoxications, 2nd Ed. Academic Press, New York. 748 pp.

Bryan, F. L. 1980. Foodborne diseases in the United States associated with meat and poultry. J. Food Prot. 43:140-150.

Center for Disease Control. 1979. Botulism in the United States, 1899-1977. Handbook for Epidemologists, Clinicians and Laboratory Workers. Center for Disease Control, U.S. Department of Health, Education, and Welfare, Public Health Service, Atlanta, Georgia. 41 pp.

Genigeorgis, C., and H. Riemann. 1979. Food processing and hygiene. Pp. 613-713 in H. Riemann and F. L. Bryan, eds. Food-Borne Infections and Intoxications, 2nd Ed. Academic Press, New York. 748 pp.

Hauschild, A. H. W., and R. Hilsheimer. 1980. Incidence of Clostridium botulinum in commercial bacon. J. Food Prot. 43:564-565.

Holley, R. A. 1978. Botulism Hazard in Cured Meats Treated With Reduced Concentrations of Nitrite: A Report Prepared for the Industry/Government Committee on Nitrites and Nitrosamines, Food Research Institute, Research Branch, Agriculture Canada, Ottawa, Ontario, Canada. 20 pp.

International Commission on Microbiological Specifications for Foods. 1980. Microbial Ecology of Foods. Volumes 1 (pp. 1-332) and 2 (pp. 333-998). Academic Press, New York.

Johnston, R. W., and R. P. Elliott. 1976. Meat and poultry products. Pp. 540-548 in M. L. Speck, ed. Compendium of Methods for the Microbiological Examination of Foods. American Public Health Association, Washington, D.C.

Morris, J. G., and C. L. Hatheway. 1980. Botulism in the United States, 1979. J. Infect. Dis. 142:302-305.

National Academy of Sciences. 1969. An Evaluation of the Salmonella Problem. National Academy of Sciences, Washington, D.C. 207 pp.

National Academy of Sciences. 1975. Prevention of Microbial and Parasitic Hazards Associated with Processed Foods. A Guide for the Food Processor. National Academy of Sciences, Washington, D.C. 166 pp.

National Academy of Sciences. 1981. The Health Effects of Nitrate, Nitrite, and *N*-Nitroso Compounds. National Academy Press, Washington, D.C. 544 pp.

Roberts, T. A., and J. L. Smart. 1976. The occurrence and growth of *Clostridium* spp. in vacuum-packed bacon with particular reference to *Cl. perfringens* (*welchii*) and *Cl. botulinum*. J. Food Technol. 11:229-244.

Roberts, T. A., A. M. Gibson, and A. Robinson. 1981a. Factors controlling the growth of *Clostridium botulinum* types A and B in pasteurized, cured meats. I. Growth in pork slurries prepared from "low" pH meat (pH range 5.5-6.3). J. Food Technol. 16:239-266.

Roberts, T. A., A. M. Gibson, and A. Robinson. 1981b. Factors controlling the growth of *Clostridium botulinum* types A and B in pasteurized, cured meats. II. Growth in pork slurries prepared from "high" pH meat (pH range 6.3-6.8). J. Food Technol. 16:267-281.

Roberts, T. A., A. M. Gibson, and A. Robinson. 1981c. Prediction of toxin production by *Clostridium botulinum* in pasteurized pork slurry. J. Food Technol. 16:337-355.

Sakaguchi, G. 1979. Botulism. Pp. 389-442 in H. Riemann and F. L. Bryan, eds. Food-Borne Infections and Intoxications, 2nd Ed. Academic Press, New York. 748 pp.

Smith, L. DS. 1977. Botulism: The Organism, Its Toxins, The Disease. Charles C Thomas, Springfield, Illinois. 236 pp.

Terrell, R. N. 1974. Theory and practice of preblending for use in comminuted meat food mixes. Pp. 23-33 in Proceedings of Meat Industry Foundation Research Conference. American Meat Institute, Washington, D.C.

Tompkin, R. B. 1980. Botulism from meat and poultry products--A historical perspective. Food Technol. 34:229-236, 257.

U.S. Department of Agriculture. 1977. The Staphylococcal Enterotoxin Problem in Fermented Sausages. Food Safety and Quality Service, U.S. Department of Agriculture, Washington, D.C. 13 pp. plus appendixes.

CHAPTER 4

ANTIMICROBIAL ALTERNATIVES TO NITRITE: CHEMICAL AGENTS

Page

CHAPTER 4

ANTIMICROBIAL ALTERNATIVES TO NITRITE: CHEMICAL AGENTS

Two chapters of this report present the Committee's evaluation of antimicrobial alternatives to nitrite. In this chapter, chemical agents that control microbial proliferation are discussed. Chapter 5 focuses on the inhibitory effects of physical treatments or alterations in the intrinsic characteristics or environment of products.

Various antimicrobial compounds are used in foods. This chapter briefly reviews the major categories, focusing on their potential value for inhibiting clostridia, particularly *C. botulinum*, in meat products. For more general descriptions of the use of preservatives, the reader should consult Branen and Davidson (in press) or the International Commission on Microbiological Specifications for Foods (1980).

General reviews of the usefulness of preservatives as alternatives to nitrite have recently been published (Pierson and Smoot, in press; Sofos and Busta, 1980). No scheme for classifying preservatives is entirely satisfactory; the one used in this chapter is adopted for convenience. Many preservatives are salts of weak mineral acids (e.g., nitrites and sulfites), organic acids (e.g., acetic, sorbic, and benzoic), or salts or esters of organic acids (e.g., parabens and alkyl esters of *p*-hydroxybenzoic acid). Some antibiotics, such as nisin, have found use as preservatives. Other agents, such as sodium chloride and *S*-nitrosocysteine, are difficult to categorize.

Because the use of lower concentrations of nitrite is also a possible alternative to the conventional use of this agent, the effects of lower nitrite concentrations are assessed in this chapter.

Following a brief discussion of the methods used by the Committee to evaluate antimicrobial alternatives, the individual evaluations for each agent are presented. These evaluations focus on the efficacy against *C. botulinum*, although data on other microorganisms are also discussed. Efficacy is compared with that of nitrite used at conventional concentrations. The influence of product characteristics on the efficacy of an alternative is also discussed, along with data on its mechanism of inhibitory action. Conclusions and recommendations based on the Committee's evaluation are given at the end of the chapter.

METHODS FOR EVALUATING ALTERNATIVES

Nitrite has a multiplicity of antimicrobial effects. In its evaluation of alternatives, the Committee chose to focus on the antibotulinal effects, for two reasons: the most serious health hazard originates from spores of *C. botulinum*, and this organism can serve as a representative of other spore-forming organisms that are the major cause of spoilage in cured meats.

To compare efficacy of possible alternatives with that of nitrite, the Committee sought studies from which data could be extracted on the time required for a particular percentage of packages (inoculated with *C. botulinum* spores) to become toxic. The Committee judged that that measurement was the most reliable for comparison. Furthermore, the Committee focused on data generated with products made under commercial or pilot-plant conditions, because (as discussed in Chapter 11) such data are more reliable than laboratory data in assessing the feasibility of proposed alternatives.

Data on the time required for development of toxicity in a particular percentage of packages facilitate comparison of efficacy in two ways: they enable the delay produced (i.e., the protection provided) by an agent to be easily comprehended (e.g., as extra days of resistance to toxigenesis), and they enable the probability of toxigenesis from inoculated spores to be calculated for particular combinations of product and treatment. Such data also facilitate determination of relative risk among products and treatments.

Not all reports of studies of nitrite or alternatives examined by the Committee contained data that could be expressed in the way the Committee desired. The problems in comparing studies arise for a number of reasons:

- different test organisms were used, such as *C. sporogenes*;
- different end points were recorded, such as the time to first toxic package or the time to first swollen package (which is not reliable in predicting toxin formation); or
- different test systems were used, such as meat slurries, as opposed to commercial products.

Because of the need to generate information on the antimicrobial spectrum and efficacy of alternatives in the range of products in which they might be used and on the factors that influence their activities, no single experimental design could be ideal. The Committee discussed in detail (see Chapter 11) some of the above-mentioned experimental considerations that it believes important. A brief bibliography of studies that were conducted in model systems and are not discussed in this chapter is presented in Appendix A.

EVALUATION OF ALTERNATIVES

Low Concentrations of Nitrite

The effects of conventional nitrite use were reviewed by the Committee in its first report (National Academy of Sciences, 1981). The antibotulinal effects of nitrite at concentrations below those currently used are reviewed here. Table 4-1 present some selected examples of the

TABLE 4-1

Delay by Nitrite of Toxin Production or Swelling in Various Commercial Cured Products Inoculated with *C. botulinum* Spores and Subjected to Simulated Temperature Abuse[a]

Product Class	Product	Inoculum Spores/g[b]	Processing[c]	Sodium Chloride, %	Days to Comparable Toxin Production or Swelling, Ratio of No Nitrite to: 30 mg/kg	40 mg/kg	50 mg/kg	60 mg/kg	100 mg/kg	120 mg/kg	150 mg/kg	156 mg/kg	200 mg/kg	Samples Toxic or Swollen at Time of Comparison, %	References
Raw cured															
Bacon	Bacon	52	--[d]	1.33	∿10:∿10	--	--	∿10:∿20	--	∿10:56	—	--	--	5	Christiansen *et al.*, 1974
Fermented sausage	Summer sausage[e]	100	58.3	2.5	--	--	<14:<14	--	--	--	<14:<21	--	--	33	Christiansen *et al.*, 1975
Cooked cured															
Pasteurized, packaged after heating	Turkey ham	5	68.3	1.07	--	--	<3:<6	--	<3:<10	--	--	<3:<12	--	80	Christiansen *et al.*, 1977
	Frankfurters, meat	325	71	2.6	--	--	<14:<56	--	--	--	--	<14:>56	--	20	Hustad *et al.*, 1973
	Frankfurters, chicken	5	--	--	--	--	<7:<6	--	<7:>12	--	--	<7:>12	--	80	Christiansen *et al.*, 1977
	Frankfurters, chicken	500	68.5	2.5	--	<4:<4	--	--	--	--	--	<4:<6	--	50	Sofos *et al.*, 1979d
Pasteurized, canned	Perishable, canned comminuted pork	100	68.5	2.5	--	--	7:43	--	7:102	--	--	7:107	--	50	Tompkin, 1978; Tompkin *et al.*, 1977
Shelf-stable	Pork luncheon meat	100	0.4[f]	2.3	--	--	--	--	10:14	--	--	--	10:17	50	Pivnick and Chang, 1973

[a]The efficacy of various treatments is analyzed by comparing the number of days of incubation that were required for a particular percentage of products in a treatment group to become toxic or swollen at the simulated abuse (incubation) temperatures. The incubation temperature was 27°C in all studies except that of Pivnick and Chang (1973), which was 35°C. Bacon used in the study by Christiansen *et al.* (1974) was produced in a pilot plant.

[b]Data in the column represent the lowest spore inoculum used in the cited studies.

[c]Expressed as center temperature, °C.

[d]--Denotes not tested or not recorded.

[e]Data presented are for products formulated without glucose and starter culture. For products to which glucose and starter culture, but no nitrite, had been added, 10% of the samples were toxic within 112 d. For products to which glucose and sodium nitrite (50 mg/kg), but no starter culture, had been added, toxin production was completely inhibited as long as 112 d.

[f]Expressed as lethal value (F_o). A botulinum cook has an F_o value of 2.78.

delay in toxin production or swelling in products inoculated with *C. botulinum* spores and subjected to simulated temperature abuse. These data demonstrate that some products, if formulated with no nitrite, would become toxic before others--e.g., turkey ham before perishable canned comminuted pork or red-meat frankfurters, and the protection (e.g., delay of toxigenesis) conferred by a specific concentration of added nitrite varies among products.

For bacon, no discernible difference was seen until nitrite was added at 60 mg/kg, and a concentration of 120 mg/kg was required to produce a substantial difference (Christiansen *et al.*, 1974). A concentration of 150 mg/kg was required before inhibition was detected in summer sausage (Christiansen, *et al.*, 1975). However, differences were seen at 50 mg/kg in turkey ham, meat frankfurters, and perishable canned comminuted pork (Christiansen *et al.*, 1977; Hustad *et al.*, 1973; Tompkin, 1978; Tompkin *et al.*, 1977).

Appendix D presents calculations of the probability of toxigenesis or swelling from *C. botulinum* spores in products with different amounts of nitrite under conditions of simulated temperature abuse. Results with bacon (Table D-1), perishable canned comminuted pork (Table D-2), and some other products (Table D-3) are summarized in Table 4-2.

For bacon, on which most testing has been done, the probabilities of toxigenesis in a nitrite-free product subjected to a temperature of 27°C for a week variees by 6 orders of magnitude, depending on a variety of intrinsic product factors, such as sugar, salt, ascorbate, isoascorbate, and phosphate concentrations (Chapter 3). For most concentrations of added sodium nitrite, the lower probabilities of toxigenesis seem to be usually, but not invariably, associated with higher sugar or brine concentrations. Insufficient data are available to judge accurately the effect of ascorbate on the probability of toxigenesis in low-nitrite bacon.

Within particular studies, it is apparent that addition of increasing concentrations of nitrite generally reduces the probability of toxigenesis from *C. botulinum* spores. Taken overall, these data reinforce the view that intrinsic product factors (other than nitrite)--particularly added carbohydrate and salt--and factors intrinsic to the raw meat before processing can considerably increase the resistance of bacon to toxigenesis.

Because of the wide variation in the intrinsic susceptibility (or resistance) to toxigenesis from *C. botulinum* spores during temperature abuse of nitrite-free products, as well as among supposedly similar products (e.g., bacon), omission of nitrite from different classes of product would have different results. Product classes differ in the increase in resistance to toxigenesis that is conferred by raising the concentration of nitrite; for example, fermented sausages display a

TABLE 4-2

Effect of Nitrite on the Probability of Outgrowth and Toxigenesis or Swelling from *C. botulinum* Spores During Simulated Temperature Abuse for 1 Week at 27°C in Various Cured-Meat Products[a]

Product Class	Product	Added Nitrite Concentration, mg/kg: 0	40-50	100	120	150-156
Raw, cured						
Bacon	Bacon(s)	1×10^{-2}–$<6 \times 10^{-8}$	5×10^{-4}–6×10^{-8}	$<1 \times 10^{-5}$–1×10^{-7}	3×10^{-4}–$<2 \times 10^{-8}$	10^{-5}–$<6 \times 10^{-8}$
Other pickle-cured						
Dry-cured						
Fermented, dried and semidried sausages	Fermented, e.g., summer sausage	$\leq 10^{-5}$	$<10^{-5}$	$<10^{-5}$		$<10^{-5}$
Cooked, cured						
Pasteurized, packaged after	Liver sausage (ms)[b]	$\geq 10^{-2}$–10^{-4}	$\geq (10^{-2}$–$10^{-4})$	10^{-2}–10^{-5}		$<10^{-6}$
	Turkey ham	$\geq 10^{-2}$–10^{-4}		10^{-2}–10^{-5}		$<(10^{-3}$–$10^{-4})$
	Chicken frankfurters	$\geq 10^{-2}$–10^{-4}		$<10^{-3}$		$<(10^{-3}$–$10^{-4})$
	Frankfurters	10^{-4}–$<10^{-5}$[c]	$\leq 10^{-5}$	$<10^{-5}$		$<10^{-5}$
	Liver sausage (hs)[b]	$\leq 10^{-5}$	$<10^{-5}$	$<10^{-5}$		$<10^{-5}$
Pasteurized, canned	Comminuted pork	$>3 \times 10^{-3}$[d]	$>3 \times 10^{-3}$–$>5 \times 10^{-4}$[d]			5×10^{-4}–5×10^{-6}[d]
		$>3 \times 10^{-3}$–$>10^{-6}$[e]	8×10^{-4}–$<7 \times 10^{-7}$[e]	$<2 \times 10^{-5}$–6×10^{-7}[e]		3×10^{-5}–$<3 \times 10^{-6}$[e]
Commercially sterile	e.g., deviled ham	$<10^{-12}$	$<10^{-12}$	$<10^{-12}$	$<10^{-12}$	$<10^{-12}$

[a]Probability, $p = (\ln n/q)/s$, where n = no. packages, q = no. nontoxic packages, and s = no. challenge spores/package. Data in this table extracted from Tables D-1, D-2, and D-3 in Appendix D.

[b]ms = moderately salted, hs = highly salted.

[c]Includes some tests conducted with sodium nitrite at 30 mg/kg.

[d]Without added ascorbate or isoascorbate.

[e]With sodium ascorbate or isoascorbate at ∿200 mg/kg.

smaller increase in resistance to toxigenesis than do pasteurized poultry products for equal additions of nitrite. This is consistent with data presented in Table 4-1. Variations in susceptibility to toxigenesis can occur within product classes as well. For example, the probability of toxigenesis varies by 4 orders of magnitude in different samples of bacon containing sodium nitrite at 120 mg/kg. In this case, factors such as added carbohydrate, brine concentration, isoascorbate, and meat composition affect resistance to toxigenesis (Chapter 3). If more were known about the interactions of these factors, cure formulations could be altered to optimize microbial inhibition, and the amount of nitrite added to products could be reduced to some extent without increasing the risk of pathogenic hazard.

Because of the wide variation among supposedly similar products, it may not be possible to use calculations of probabilities of toxigenesis for other than the most general comparisons between product classes. The observed variation underscores the need for a better understanding of the interactions of the factors that contribute to control of microbial proliferation. Knowledge of interactions was summarized in the Committee's first report, which also described attempts to derive models that could be used to predict changes in protection against toxigenesis if factors that control microbial proliferation were altered (National Academy of Sciences, 1981). Perhaps the most comprehensive of such models is that derived by Roberts *et al.* (1981a,b,c), who studied such variables as nitrite, nitrate, sodium chloride, isoascorbate, heat treatment, and abuse temperature in a heated meat-slurry model system.

Other Salts of Weak Mineral Acids

Hypophosphite. Sodium hypophosphite, the sodium salt of hypophosphorous acid, has been investigated by Pierson *et al.* (1981b) as an alternative to nitrite for inhibiting *C. botulinum* in bacon produced commercially. The delay in toxin production or swelling in bacon inoculated with *C. botulinum* spores and containing sodium hypophosphite or nitrite is shown in Table 4-3. Sodium hypophosphite at 3,000 mg/kg appears to be as effective as sodium nitrite at 120 mg/kg. In addition, the combination of sodium nitrite at 40 mg/kg with sodium hypophosphite at 3,000 mg/kg, or 1,000 mg/kg in the presence of sodium tripolyphosphate, appears to be at least as effective as current nitrite use. Sodium hypophosphite has not been studied in other meat products, nor has it been assessed against microorganisms other than *C. botulinum*.

Sulfite, Sulfur Dioxide, Bisulfite, and Metabisulfite. Sulfur dioxide, or an agent that produces sulfur dioxide, is used in various countries in many products--e.g., fruits, shrimp, sausage, pickles, and wines--to control microbial growth. It inhibits molds, bacteria, and undesirable yeasts. The use of sulfur dioxide and related salts to

TABLE 4-3

Delay by Sodium Hypophosphite in Toxin Production and Swelling from *C. botulinum* Spores in Commercially Prepared Bacon Subjected to Simulated Temperature Abuse at 27°C[a]

Conditions			Inoculum, Spores/g: 100-500	Inoculum, Spores/g: 100-500	Uninoculated
Sucrose, mg/kg			1,100	1,100	1,100
Sodium tripolyphosphate, mg/kg			2,000-3,000	2,000-3,000	2,000-3,000
Sodium chloride			1.5-2.0	1.5-2.0	1.5-2.0
Isoascorbate, mg			550	500	550
	Sodium Nitrite, mg/kg	Sodium Hypophosphite, mg/kg	Ratio, Days to First Toxin		Ratio, % Swollen Packages After 17 d
No nitrite:	120	0	7:14	6:14	100:20
No nitrite:	40	0	7:12	6:7	100:50
No nitrite:	40	500	7:7	--[b]	100:80
No nitrite:	40	750	--	6:7	--
No nitrite:	40	1,000	7:37	6:15 (6:10[c])	100:20
No nitrite:	40	3,000	7:17	6:14	100:10
No nitrite:	0	500	7:9	--	100:100
No nitrite:	0	1,000	7:7	6:4	100:60
No nitrite:	0	2,000	--	6:4	--
No nitrite:	0	3,000	7:17	6:10	100:30

[a]Data from Pierson *et al.*, 1981b.
[b]--Denotes not tested or not recorded.
[c]Without sodium tripolyphosphate.

control microorganisms in foods has been reviewed by the International Commission on Microbiological Specifications for Foods (1980, pp. 180-184) and Ough (in press). The active inhibitory species is the unbound unionized molecular form that increases as pH is lowered. Inhibitory action is most pronounced below at pH of 4.

The only investigations of sulfur dioxide as a partial or complete antibotulinal replacement for nitrite appear to be those conducted by Tompkin *et al*. (1980a,b). Sodium metabisulfite was used to provide sulfur dioxide at concentrations of 50, 100, or 200 mg/kg. In addition, sulfur dioxide at 200 mg/kg was used in combination with sodium nitrite at 40 mg/kg. In perishable canned comminuted pork inoculated with *C. botulinum* spores and incubated at 27°C, the various concentrations of sulfur dioxide led to variable inhibition of swelling, and the antibotulinal effect increased dramatically as the sulfur dioxide concentration was raised from 100 mg/kg to 200 mg/kg and was essentially unaffected by storage at 4.4°C for 10 wk before temperature abuse. However, the presence of nitrite significantly reduced the antibotulinal efficacy of sulfur dioxide at 200 mg/kg.

The maintenance of antibotulinal activity during storage is a desirable feature of alternatives to nitrite. However, the apparent antagonism between sulfur dioxide and nitrite may limit its usefulness where it is desired to use nitrite for color fixation and antioxidant effects. In the United States, sulfur dioxide is not permitted in foods that are major sources of thiamine, such as meats.

Organic Acids and Their Salts

The organic acids most commonly used as food preservatives are acetic, propionic, lactic, citric, sorbic, and benzoic. Because the acid or ester must remain undissociated to inhibit microbial growth, the efficacy of these agents depends on their equilibrium dissociation constants and the pH of their environment.

Probably because their optimal pH (2.5-4.0) is well below that of most cured meats, benzoic acid and benzoates do not appear to have been tested as replacements for nitrite (Robach, 1980). Similarly, no studies on the anticlostridial efficacy of propionic acid, propionates, or citric acid in meat products were found. The optimal pH of sorbic acid, however, is 4.75, and the upper limit for significant activity may be 6.5 (Sofos and Busta, in press).

Sorbic Acid and Sorbates.[1] Of all potential alternatives to nitrite, sorbic acid (CH_3-CH=CH-CH=CH-COOH), usually as the potassium salt, has been the most intensively investigated. Studies on sorbic acid have been thoroughly reviewed by Sofos *et al.* (1979a), Robach and Sofos (in press), and Sofos and Busta (in press), and only the most pertinent findings are mentioned here. The use of sorbic acid in foods other than meats has been reviewed by the International Commission on Microbiological Specifications for Foods (1981, pp. 126-135) and Sofos and Busta (in press).

Tompkin *et al.* (1974) reported that sorbic acid delayed production of toxin from *C. botulinum* spores in an uncured, cooked sausage product. This finding led to intensive investigation of its capacity to serve as a total or partial replacement for nitrite in various products (Sofos and Busta, in press).

In 1979, the U.S. Department of Agriculture investigated the efficacy of the combination of potassium sorbate at 2,600 mg/kg) and sodium nitrite at 40 mg/kg in bacon commercially produced on a large scale. The results are shown in Table 4-4 with those of other studies of sorbic acid alone at 0.2% (from potassium sorbate at 2,600 mg/kg and in combination with sodium nitrite at 40 or 80 mg/kg. The data show that the combination of potassium sorbate at 2,600 mg/kg and sodium nitrite at 40 or 80 mg/kg provides antibotulinal protection generally equivalent to that afforded by sodium nitrite alone at 120 mg/kg. Sorbate alone at 2,600 mg/kg was less effective.

The combination of sorbic acid and nitrite has also been tested in various frankfurter emulsions and found to be as effective as conventional nitrite use, as shown in Table 4-5. In this case, as was seen in bacon, sorbic acid alone at 2,000 mg/kg is not as effective as conventional use of nitrite. However, Ivey and Robach (1978) found that, in canned comminuted perishable pork, sorbic acid at 2,000 mg/kg with sodium acid pyrophosphate at 0.5% was more effective than sodium nitrite at 156 mg/kg in inhibiting the swelling that results from *C. botulinum* spore outgrowth. Treatments with sorbic acid at 2,000 mg/kg plus sodium nitrite at 50 mg/kg or sorbic acid at 1,000 or 2,000 mg/kg plus sodium nitrite at 50 mg/kg and sodium acid pyrophosphate at 1.5% were also at least as effective as conventional use of sodium nitrite at 156 mg/kg. Huhtanen and Feinberg (1980) studied sorbic acid activity in commercially prepared frankfurters and found that 0.2% sorbic acid in turkey frankfurters acidified to a pH of 5.7 resulted in protection comparable with that of sodium nitrite at 135 mg/kg (as measured by mean swell time). This mixture was less effective in chicken frankfurters.

---

[1] The term "sorbic acid" is used here to denote both potassium sorbate and sorbic acid, because it is the undissociated form of the acid that is active.

TABLE 4-4

Delay in Toxin Production by Various Preservatives or Combinations in Commercially Produced Bacon Inoculated with 1,000 C. botulinum Spores and Subjected to Simulated Temperature Abuse at 27°C[a]

Conditions		USDA (1979)[b]	USDA (1979)[b]	USDA (1979)[b]	USDA (1979)[b]	Sofos et al. (1980c)	Sofos et al. (1980c)
Cure ingredients[c]							
Sugar, mg/kg		600	20	1,200	5,000	8,000	8,000
Phosphate, mg/kg		3,000	750	1,200	4,600	5,000	5,000
Sodium chloride, %		1.47	1.69	1.81	1.40	1.70	1.70
Days to comparable toxin production or swelling; ratio of no nitrite to:							
Sodium Nitrite, mg/kg	Potassium Sorbate, mg/kg						
120	0	5:12	10:15	6:16	15:>60	13:>60	<10:>40
80	0	--[d]	--	--	--	--	--
60	0	--	--	--	--	--	--
40	0	--	--	--	--	--	--
80	2,600	--	--	--	--	13:>60	<10:>40
40	2,600	5:33	10:23	6:30	15>60	13:55	<10:10
0	2,600	--	--	--	--	13:20	<10:10
Percent samples toxic at time of comparison		50	40	20	20	20	20

[a]All tests were conducted in bacon commercially produced on a large scale. The efficacy of various treatments is analyzed by comparing the number of days of incubation that were required for a particular percentage of packages in a treatment group to become toxic or swollen. All data except those in the last column are for toxin production in swollen packages. Data in the last column are for toxin production in serially sampled packages, whether swollen or not swollen.

[b]Data pertain to Phase I of the study by the U.S. Department of Agriculture (1979).

[c]Concentrations listed were target values. Actual values may deviate, sometimes considerably, from target values (e.g., U.S. Department of Agriculture, 1979).

[d]Denotes not tested or not recorded.

TABLE 4-5

Delay in Toxin Production by Various Preservatives or Combinations in Frankfurters Inoculated with 500 _C. botulinum_ Spores and Subjected to Simulated Temperature Abuse at 27°C[a]

Conditions		Beef/Soy (Sofos _et al._, 1979b, 1980b)	Beef (Sofos _et al._, 1979b, 1980b)	Pork (Sofos _et al._, 1979b, 1980b)	Chicken (Sofos _et al._, 1979c,d, 1980a)	Chicken (Sofos _et al._, 1979c,d, 1980a)
Days to comparable toxin production; ratio of no nitrite to:						
Sodium Nitrite, mg/kg	Potassium Sorbate, mg/kg					
156	0	5:31	--[b]	4:13	<4:6	4:6
80	0	5:4	4:8	4:6	--	4:4
40	0	--	--	--	<4:<4	--
80	2,600	5:52	4:45	4:45	--	4:17
40	2,600	--	--	--	<4:10	--
0	2,600	5:8	4:8	4:8	--	4:7
Percent samples time of comparison		20	20	20	50	20

[a]All tests were conducted in frankfurters produced under commercial conditions. The efficacy of various treatments is analyzed by comparing the number of days of incubation that were required for a particular percentage of packages in a treatment group to become toxic.

[b]Denotes not tested or not recorded.

Sorbic acid in combination with sodium nitrite has also been shown to inhibit staphylococci and salmonellae (Pierson *et al.*, 1979; Tompkin *et al.*, 1974). It does not, however, inhibit lactic-acid-producing organisms (Emard and Vaughn, 1952). The other antimicrobial actions, predominantly antifungal action, of sorbic acid in uncured meats and other foods have been reviewed by Sofos and Busta (in press).

The mechanism of action of sorbic acid has been investigated, and studies have indicated that it acts as a competitive inhibitor of spore germination (Pierson *et al.*, 1981a; Smoot and Pierson, 1981; Sofos *et al.*, 1979c). The precise mechanism by which sorbic acid is inhibitory is not fully elucidated (Sofos and Busta, in press), especially that of its apparent synergism with nitrite. Recent studies indicate that sorbic acid may also inhibit spore outgrowth or cell multiplication (J. C. Blocher and F. F. Busta, University of Minnesota, personal communication, 1981; Sofos *et al.*, 1979d).

More details on the investigations of the efficacy of sorbic acid as a total or partial replacement for nitrite can be found in Pierson and Smoot (in press), Sofos (1981), and Sofos and Busta (1980).

Esters of Organic Acids

A number of organic acid esters have been tested in meat products or systems as possible replacements for nitrite, generally with disappointing results. They are active in laboratory media or aqueous slurries, but because of their higher lipid solubility, they are generally not available for antimicrobial action in the aqueous phase of meat products.

Parabens. Tanaka *et al.* (1978) found that parabens--the alkyl esters of *p*-hydroxybenzoic acid--had little antibotulinal activity in frankfurters. Deibel (1979) reported similar findings.

Glyceryl Monolaurate. When tested in a meat slurry (pH, 6.0-6.2), glyceryl monolaurate (monolaurin; Lauricidin) at 5 g/kg inhibited toxin production by *C. botulinum* strains producing toxin type A, B, or E, but at 3 g/kg there was no activity (Notermans and Dufrenne, 1981). Robach *et al.* (1981) also reported that 3 g/kg had no effect on outgrowth and gas formation by *Clostridium sporogenes* PA 3679 in a pork homogenate. This compound has not been compared with nitrite in a model system or tested in a meat product.

Fumarate Esters. Methylfumarate and ethylfumarate esters at 1,500 or 2,000 mg/kg were reported to be at least as effective in inhibiting gas production in a comminuted bacon model system inoculated with *C. botulinum* spores as sodium nitrite at 120 mg/kg (Huhtanen *et al.*, 1981). In a small-scale study, bacon produced under simulated commercial conditions with monomethylfumarate or monoethylfumarate at 1,250 mg/kg has been found to be at least as resistant to toxigenesis from *C. botulinum* spores as conventional bacon (produced with sodium nitrite at 120 mg/kg) (Huhtanen, *et al.*, 1981; U.S. Department of Agriculture Eastern Regional Research Center, Philadelphia, personal communication, 1981).

Antibiotics

The use of antibiotics as antimicrobial food additives has been reviewed by the International Commission on Microbiological Specifications for Foods (1981, pp. 160-169). In this country, largely because of concerns regarding the selective pressures for antibiotic resistance that they would impose, antibiotics are not used as food additives; however, two nontherapeutic antibiotics are used in foods in other countries--natamycin and nisin. Because natamycin is effective primarily against fungi, it is not discussed here.

The use of nisin as an antimicrobial adjunct to heat processing of some foods and as a means of controlling the clostridial "blowing" of cheese has been thoroughly reviewed by Lipinska (1977); however, testing of nisin as an alternative to nitrite in cured meats has been sparse. Rayman *et al.* (1981) reported that nisin at 75 mg/kg was superior to sodium nitrite at 150 mg/kg in inhibiting outgrowth of *C. sporogenes* PA 3679 in meat slurries that had been processed to simulate the cooking of ham. However, during refrigerated storage, nisin depleted to a point where it no longer inhibited clostridial outgrowth during temperature abuse. A combination of nisin at 75 or 100 mg/kg and sodium nitrite at 40 mg/kg almost completely inhibited outgrowth after refrigerated storage. Later studies of the effect of nisin in model meat systems have indicated that it is less active against *C. botulinum* spores than against *C. sporogenes* spores (A. Hurst, Health and Welfare Canada, Ottawa, Ontario, personal communication, 1981).

S-Nitrosocysteine

Kanner and Juven (1980) investigated *S*-nitrosocysteine, which is generated by nitrite during the curing of meat, as an anticlostridial agent. It was less effective than nitrite at delaying swelling of perishable comminuted canned turkey meat that had been inoculated with *C. sporogenes* PA 3679 spores, heated to 68.5°C, and then incubated at 27°C. Fifty percent of cans containing 1.45 mM *S*-nitrosocysteine were swollen after 14 d of incubation. Doubling the *S*-nitrosocysteine concentration postponed this end point by only 2 d. However, swelling of half the cans containing 1.45 mM sodium nitrite (100 mg/kg) was not observed until after 23 d of incubation; if the nitrite concentration was doubled, the end point was postponed to over 100 d. No data were collected in this study on swelling of cans containing neither nitrite nor *S*-nitrosocysteine.

Sodium Chloride

Sodium chloride is an indispensable agent in curing and in the system contributing to the preservation of meat, poultry, and fish products to which nitrite is added. It acts to control microbial proliferation, not only through the lowering of water activity (Chapter 5), but apparently also through its own antimicrobial action (Baird-Parker and Freame, 1967). The concentration of sodium chloride in the aqueous phase--the brine concentration--is important in selective inhibition of microbial growth in cured products.

The effects of sodium chloride on botulinal hazard and microbial spoilage and the range of microorganisms affected have been reviewed elsewhere (International Commission on Microbiological Specifications for Foods, 1980; National Academy of Sciences, 1981). Brine concentrations of over 8-10% inhibit spore outgrowth of Group I *C. botulinum* (e.g., proteolytic strains producing toxin type A or B), whereas outgrowth of spores of Group II (e.g., nonproteolytic strains producing toxin type E or B) is limited by brine at 5% or greater. Increasing the brine concentration of cured products could reduce the need for nitrite in those products; however, the extent to which this could be done is limited, not so much by considerations of antimicrobial efficacy as by the impact of such an increase on the sensory characteristics (acceptability) of different products and the toxicologic consequences of increased sodium intake. These issues are addressed in Chapters 9 and 10.

CONCLUSIONS

Adequate assessment of the usefulness of anticlostridial alternatives to nitrite requires that testing be conducted in products made on a large scale under a variety of commercial conditions. Under these conditions, antimicrobial alternatives to nitrite should, at a minimum, have equal antibotulinal activity and preferably activity equal to that of conventional concentrations of nitrite against spoilage clostridia and other spore-forming microorganisms of concern in cured meats. Activity against staphylococci (particularly *Staphylococcus aureus*) and salmonellae, against which nitrite has little or no activity, would be a desirable feature of alternatives. Similarly, it is desirable that alternatives be less inhibitory toward lactic-acid-producing bacteria than to pathogens, because the former may ferment added carbohydrate to acid during temperature abuse, and that would result in a lowered pH, which inhibits development of pathogens.

Omission of nitrite in cured products had varied effects on antibotulinal protection in different classes of cured products. In addition, the effect of particular concentrations of nitrite on antibotulinal protection within product categories also varied greatly. These findings underscore the importance of other intrinsic product factors (Chapter 3) in inhibiting *C. botulinum* toxigenesis. If cure formulations were altered to maximize microbial inhibition, nitrite concentrations could possibly be reduced; however, more information on how to optimize inhibition by intrinsic product factors is needed before such a reduction in nitrite concentrations would be possible.

Of the chemical inhibitors studied thus far as alternatives to nitrite, only the combination of sorbic acid (potassium sorbate at 2,600 mg/kg) and sodium nitrite (at 40-80 mg/kg) has been convincingly demonstrated to display antibotulinal activity equivalent to that of conventional nitrite use in commercially processed bacon. In addition, this combination is effective in frankfurters, and sorbic acid and sodium acid pyrophosphate with or without nitrite at 50 mg/kg was as effective in inhibiting *C. botulinum* outgrowth as nitrite alone at 156 mg/kg in canned comminuted perishable pork. Sorbic acid inhibits salmonellae and *Staphylococcus aureus*, and some information is available on its mode of action and on factors that influence its activity.

Sodium hypophosphite, alone at 3,000 mg/kg or at 1,000 or 3,000 mg/kg in combination with sodium nitrite at 40 mg/kg, has displayed antibotulinal activity equivalent to that of conventional nitrite use in bacon that was produced in relatively small amounts in two commercial-plant studies. Investigations with a comminuted-bacon model system and preliminary reports on bacon produced on a small scale under conditions simulating commercial production have indicated that some fumarate esters at 1,250 mg/kg show antibotulinal activity equal to that of conventional nitrite use. However, more information on the activity of sodium hypophosphite and methylfumarate in various products, their mechanisms of action, and their toxicity is needed before decisions can be made on their introduction as alternatives to nitrite.

Increased additions of sodium chloride could be used to achieve the desired inhibition of microbial growth, but this would need to be substantial, especially to inhibit *Staphylococcus aureus*.

No other chemical inhibitors of microbial growth can yet be judged as promising replacements for nitrite.

RECOMMENDATIONS

Studies should be undertaken to determine how to optimize control of microbial proliferation by intrinsic product factors other than nitrite (Chapter 3). Information derived from such studies should enable producers to assess the possibility of reducing added nitrite while retaining the desired degree of microbial inhibition through adjustment of controlling factors other than nitrite.

The results obtained with combinations of nitrite and sorbic acid in frankfurters and comminuted perishable pork should be verified and extended to other products made under a variety of commercial production conditions.

The antibotulinal activity displayed by sodium hypophosphite and fumarate esters in bacon should be verified, and tests should be conducted in other products.

If an alternative is found to display antibotulinal activity equivalent to that of conventional nitrite use, its activity against other pathogens should be investigated in conjunction with studies of factors that influence its activity, as described for nitrite.

REFERENCES

Baird-Parker, A. C., and B. Freame. 1967. Combined effect of water activity, pH and temperature on the growth of *Clostridium botulinum* from spore and vegetative cell inocula. J. Appl. Bacteriol. 30:420-429.

Branen, A. L., and P. M. Davidson, eds. In press. Antimicrobials in Foods. Marcel Dekker Co., New York.

Christiansen, L. N., R. B. Tompkin, A. B. Shaparis, T. V. Keuper, R. W. Johnston, D. A. Kautter, and O. J. Kolari. 1974. Effect of sodium nitrite on toxin production by *Clostridium botulinum* in bacon. Appl. Microbiol. 27:733-737.

Christiansen, L. N., R. B. Tompkin, A. B. Shaparis, R. W. Johnston, and D. A. Kautter. 1975. Effect of sodium nitrite and nitrate on *Clostridium botulinum* growth and toxin production in a summer style sausage. J. Food Sci. 40:488-490.

Christiansen, L. N., J. C. Manuel, and L. S. Divilbiss. 1977. Effect on sodium nitrite in cured poultry products. In Special Poultry Research Committee, Response to FDA Notice Published in September 2, 1977, *Federal Register* (Fed. Reg. 42:44376; Docket No. 77N-0222).

Deibel, R. H. 1979. Parabens. Presented at the Meat Industry Research Conference, March 29-30, 1979, American Meat Institute, Washington, D.C.

Emard, L. O., and R. H. Vaughn. 1952. Selectivity of sorbic acid media for the catalase negative lactic acid bacteria and clostridia. J. Bacteriol. 63:487-494.

Huhtanen, C. N., and J. Feinberg. 1980. Sorbic acid inhibition of *Clostridium botulinum* in nitrite-free poultry frankfurters. J. Food Sci. 45:453-457.

Huhtanen, C. N., M. Dymicky, and H. Trenchard. 1981. Methyl and ethyl esters of fumaric acid as substitutes for nitrite for inhibiting *Clostridium botulinum* spore outgrowth in bacon. Presented at the 41st Annual Meeting of the Institute of Food Technologists, June 7-10, 1981, Atlanta, Georgia. Abstract 21.

Hustad, G. O., J. G. Cerveny, H. Trenk, R. H. Deibel, D. A. Kautter, T. Fazio, R. W. Johnston, and O. E. Kolari. 1973. Effect of sodium nitrite and sodium nitrate or botulinum toxin production and nitrosamine formation in wieners. Appl. Microbiol. 26:22-26.

International Commission on Microbiological Specifications for Foods. 1980. Microbial Ecology of Foods. Volumes 1 (pp. 1-332), and 2 (pp. 333-998). Academic Press, New York.

Ivey, F. J., and M. C. Robach. 1978. Effect of sorbic acid and sodium nitrite on Clostridium botulinum outgrowth and toxin production in canned comminuted pork. J. Food Sci. 43:1782-1785.

Kanner, J., and B. J. Juven. 1980. S-Nitrosocysteine as an antioxidant, color-developing, and anticlostridial agent in comminuted turkey meat. J. Food Sci. 45:1105-1112.

Lipinska, E. 1977. Nisin and its applications. Pp. 103-130 in M. Woodbine, ed. Antibiotics and Antibiosis in Agriculture. Butterworths Publishers Inc., Boston, Massachusetts.

National Academy of Sciences. 1981. The Health Effects of Nitrate, Nitrite, and N-Nitroso Compounds. National Academy Press, Washington, D.C. 544 pp.

Notermans, S., and J. Dufrenne. 1981. The effect of glyceryl monolaurate on toxin production by Clostridium botulinum in meat slurry. J. Food Safety 2:83-88.

Ough, C. S. In press. Sulfur dioxide and sulfite. In A. L. Branen and P. M. Davidson, eds. Antimicrobials in Foods. Marcel Dekker, New York.

Pierson, M. D., and L. A. Smoot. In press. Nitrite, nitrite alternatives and the control of Clostridium botulinum in cured meats. CRC Crit. Rev. Food Sci. Nutr.

Pierson, M. D., L. A. Smoot, and N. J. Stern. 1979. Effect of potassium sorbate on growth of Staphylococcus aureus in bacon. J. Food Prot., 42:302-304.

Pierson, M. D., L. A. Smoot, and R. V. Lechowich. 1981a. The effect of sorbate-nitrite interactions on the inhibition of Bacillus cereus spore germination. Presented at the 81st Annual Meeting of the American Society for Microbiology, March 1-6, 1981, Dallas, Texas. Abstract, p. 20.

Pierson, M. D., K. M. Rice, and J. F. Jadlocki. 1981b. Sodium hypophosphite inhibition of Clostridium botulinum in bacon. Pp. 651-654 in Proceedings of the 27th European Meat Research Workers Congress, Volume 2, August 24-28, 1981, Vienna, Austria.

Pivnick, H., and P-C. Chang. 1973. Perigo effect in pork. Pp. 111-116 in B. Krol and B. J. Tinbergen, eds. Proceedings of the International Symposium on Nitrite in Meat Products, Zeist, PUDOC, Wageningen, the Netherlands.

Rayman, M. K., B. Aris, and A. Hurst. 1981. Nisin: A possible alternative or adjunct to nitrite in the preservation of meats. Appl. Environ. Microbiol. 41:375-380.

Rice, K. M., and M. D. Pierson. 1981. Inhibition of Salmonellae by sodium nitrite and potassium sorbate in frankfurters. Presented at the 41st Annual Meeting of the Institute of Food Technologists, June 7-10, 1981, Atlanta, Georgia. Abstract 251.

Robach, M. C. 1980. Use of preservatives to control microorganisms in Food. Food Technol. 34:81-84.

Robach, M. C., and J. N. Sofos. In press. Use of sorbates in meat products, fresh poultry and poultry products: A review. J. Food Prot.

Robach, M. C., C. S. Hickey, and E. C. To. 1981. Comparison of antimicrobial actions of monolaurin and sorbic acid. J. Food Safety 3:89-98.

Roberts, T. A., A. M. Gibson, and A. Robinson. 1981a. Factors controlling the growth of Clostridium botulinum types A and B in pasteurized, cured meats. I. Growth in pork slurries prepared from "low" pH meat (pH range 5.5-6.3). J. Food Technol. 16:239-266.

Roberts, T. A., A. M. Gibson, and A. Robinson. 1981b. Factors controlling the growth of Clostridium botulinum types A and B in pasteurized, cured meats. II. Growth in pork slurries prepared from "high" pH meat (pH range 6.3-6.8). J. Food Technol. 16:267-281.

Roberts, T. A., A. M. Gibson, and A. Robinson. 1981c. Prediction of toxin production by Clostridium botulinum in pasteurized pork slurry. J. Food Technol. 16:337-355.

Smoot, L. A., and M. D. Pierson. 1981. Mechanism of sorbate inhibition of Bacillus cereus and Clostridium botulinum 62A spore germination. Appl. Environ. Microbiol. 42:477-483.

Sofos, J. N. 1981. Nitrite, sorbate and pH interaction in cured meat products. Proceedings of the 34th Annual Reciprocal Conference of the American Meat Sciences Association, June 21-24, 1981, Corvallis, Oregon. American Meat Sciences Association, Chicago, Illinois.

Sofos, J. N., and F. F. Busta. 1980. Alternatives to the use of nitrite as an antibotulinal agent. Food Technol. 34:244-251.

Sofos, J. N., and F. F. Busta. In press. Sorbates. In A. L. Branen and P. N. Davidson, eds. Antimicrobials in Foods. Marcel Dekker, New York.

Sofos, J. N., F. F. Busta, and C. E. Allen. 1979a. Botulism control by nitrite and sorbate in cured meats. A review. J. Food Prot. 42:739-770.

Sofos, J. N., F. F. Busta, and C. E. Allen. 1979b. *Clostridium botulinum* control by sodium nitrite and sorbic acid in various meat and soy protein formulations. J. Food Sci. 44:1662-1667.

Sofos, J. N., F. F. Busta, and C. E. Allen. 1979c. Sodium nitrite and sorbic acid effects on *Clostridium botulinum* spore germination and total microbial growth in chicken frankfurter emulsions during temperature abuse. Appl. Environ. Microbiol. 37:1103-1109.

Sofos, J. N., F. F. Busta, K. Bhothipaksa, and C. E. Allen. 1979d. Sodium nitrite and sorbic acid effects on *Clostridium botulinum* toxin formation in chicken frankfurter-type emulsions. J. Food Sci. 44:668-672.

Sofos, J. N., F. F. Busta, and C. E. Allen. 1980a. Influence of pH on *Clostridium botulinum* control by sodium nitrite and sorbic acid in chicken emulsions. J. Food Sci. 45:7-12.

Sofos, J. N., F. F. Busta, and C. E. Allen. 1980b. Use of nitrite and sorbate for botulism control. Pp. 241-244 in Proceedings of the 26th European Meat Research Workers Congress, Volume 2, August 31-September 5, 1980, Colorado Springs, Colorado.

Sofos, J. N., F. F. Busta, K. Bhothipaksa, C. E. Allen, M. C. Robach, and M. W. Paquette. 1980c. Effects of various concentrations of sodium nitrite and potassium sorbate on *Clostridium botulinum* toxin production in commercially prepared bacon. J. Food Sci. 45:1285-1292.

Tanaka, N., J. G. Cerveny, N. J. Worley, E. W. Wall, and J. M. Goepfert. 1978. Effect of nitrite, sorbate, polyphosphates and parabens on toxic production by *Clostridium botulinum* in wieners. Pp. 354-358 in Annual Report, Food Research Institute, University of Wisconsin, Madison.

Tompkin, R. B. 1978. The role and mechanism of the inhibition of *C. botulinum* by nitrite--is a replacement available? Pp. 135-147 in Proceedings of the 31st Annual Reciprocal Conference of the American Meat Sciences Association, June 18-22, 1978, University of Connecticut, Storrs, Connecticut. American Meat Sciences Association, Chicago, Illinois.

Tompkin, R. B., L. N. Christiansen, A. B. Shaparis, and H. Bohn. 1974. Effect of potassium sorbate on salmonellae, *Staphylococcus aureus*, *Clostridium perfringens* and *Clostridium botulinum* in cooked, uncured sausage. Appl. Microbiol. 28:262-264.

Tompkin, R. B., L. N. Christiansen, and A. B. Shaparis. 1977. Variation in inhibition of C. botulinum by nitrite in perishable canned comminuted meats. J. Food Sci. 42:1046-1048.

Tompkin, R. B., L. N. Christiansen, and A. B. Shaparis. 1980a. Anti-botulinal efficacy of sulfur dioxide in meat. Appl. Environ. Microbiol. 39:1096-1099.

Tompkin, R. B., L. N. Christiansen, and A. B. Shaparis. 1980b. Effect of sulfur dioxide on C. botulinum in cooked meat. Pp. 248-249 in Proceedings of the 26th European Meat Research Workers Congress, Volume 2, August 31-September 5, 1980, Colorado Springs, Colorado.

U.S. Department of Agriculture. 1979. Food Safety and Quality Service Four Plant Study to Investigate the Use of 40 ppm Sodium Nitrite and 0.26 Percent Potassium Sorbate in Bacon. Food Safety and Quality Service, U.S. Department of Agriculture, Washington, D.C. 54 pp.

CHAPTER 5

ANTIMICROBIAL ALTERNATIVES: PHYSICAL TREATMENTS OR ALTERATION OF PRODUCT CHARACTERISTICS

Page

CHAPTER 5

ANTIMICROBIAL ALTERNATIVES: PHYSICAL TREATMENTS OR ALTERATION OF PRODUCT CHARACTERISTICS

A variety of physical treatments can be used to control proliferation of the microorganisms of concern in meat products. In addition, intrinsic characteristics of the product or its environment can be altered to increase microbial inhibition. These methods can be divided into three general categories:

- Application of inactivating treatments.
- Manipulation of the chemical or physical characteristics of the product.
- Maintenance of controlled environments.

For some cured-meat products, combinations of methods from all three categories currently contribute to the overall reduction in microbial proliferation. Each method has advantages and disadvantages. Some can be used alone to exert adequate control in some products. However, because the methods usually achieve the desired microbial inhibition in concert, the interactions among them are emphasized in the following discussion. In a number of instances, the efficacy of a particular method or its interactions with other product characteristics have been revealed by investigations in laboatory media. In many cases, it is not known whether these findings are pertinent to commercial meat products. Not only are the chemical and physical characteristics of food products--e.g., water activity (a_w), pH, and redox (oxidation-reduction) potential (E_h)--difficult to measure accurately, but they may vary within a product, because of the nonhomogeneity of composition of both primal and coarsely comminuted products and because of variations among manufacturers. In addition, microbial responses to treatments or adverse conditions vary widely among genera, species, and even strains within a species. Thus, studies with a particular microorganism may not be directly applicable to other microorganisms of concern.

APPLICATION OF POTENTIALLY INACTIVATING TREATMENTS

Of treatments that may inactivate microorganisms, only some types of radiation (including γ rays) and heating are practicable for food products. Extreme pH and ultraviolet radiation, for example, are not of use, because of the difficulties in applying them to foods.

Irradiation

The use of irradiation, one of the most studied treatments in food processing, has been suggested by Rowley and Brynjolfsson (1980) as a method for reducing the amount of sodium nitrite necessary to control *Clostridium botulinum*. This is only one of many potential uses of irradiation (International Commission on Microbiological Specifications for Foods, 1980, pp. 46-69). Because the advantages and disadvantages of the irradiation of food have been adequately covered elsewhere (Ingram and Farkas, 1977; International Commission on Microbiological Specifications for Foods, 1980, pp. 46-69; Rowley *et al*., 1978), the details of radappertization (i.e., radiation sufficient for commercial sterility, 4 Mrads), radurization (radiation sufficient to extend storage life, 0.5 Mrad), radicidation (radiation sufficient to remove a particular pathogen), microbial resistance, and process control will not be covered in this chapter.

Prototype radappertization processes at $-30 \pm 10^{\circ}C$, with *C. botulinum* spores types A and B as the indicator of microbiologic safety, have been carried out for such cured foods as corned beef (Anellis *et al*., 1972) and ham prepared with low concentrations of added sodium nitrite (25 mg/kg) and sodium nitrate (100 mg/kg) (Anellis *et al*., 1977). The microbiologic safety of this process is based on the dose required to reduce the number of *C. botulinum* spores by a factor of 10^{12} (Anellis *et al*., 1979). All foods irradiated at 12D have been bacteriologically sterile.* This dose would also eliminate filamentous fungi, yeast, and trichinae. Thus, highly acceptable packaged corned beef and low-concentration nitrite-nitrate ham may be produced and stored at ambient temperatures without concern for microbial spoilage or hazard to health.

Recently, Rowley *et al*. (in press) demonstrated that irradiation (at $5^{\circ}C$) provides an alternative to high concentrations of nitrite in bacon for microbiologic stability and safety. With 1.0 and 1.5 Mrads, there was a greater delay in the onset of toxic spoilage of bacon that had been inoculated with *C. botulinum* spores when irradiation was used in combination with sodium nitrite at 40 mg/kg of bacon than without nitrite. Bacon prepared with sodium nitrite added at 40 mg/kg, inoculated with approximately 2 spores/g, and irradiated with 1 Mrad remained nontoxic during 60 d at $27^{\circ}C$. About 73% of the pouches of nonirradiated bacon prepared with sodium nitrite added at 120 mg/kg and similarly treated were toxic by 18 d.

Irradiation technology is well developed, to the point of producing acceptable foods that are microbiologically safe and nonradioactive. The major concern regarding safety has been the lack of

---

*D is the dose required for 90% inactivation or one $\log_{10}$ reduction.

adequate information on the chemical products of irradiation. The toxicology of irradiated foods is discussed in Chapter 10.

Heat Treatment

The effects of high temperatures on microorganisms and the factors that influence its effects--e.g., pH, sodium chloride, carbohydrates, fats, and sodium nitrite concentrations--have recently been reviewed by the International Commission on Microbiological Specifications for Foods (1980, pp. 1-37, 136-159). Practical aspects of the use of high temperature to control microorganisms in foods have been dealt with in detail by Stumbo (1973). In its first report (National Academy of Sciences, 1981), the Committee briefly evaluated the use of heating in the control of *C. botulinum* and other organisms in cured meats. References to more detailed evaluations can be found in these publications.

Pasteurizing heat treatments generally produce temperatures at the center of products not higher than 75°C. They are effective in inactivating vegetative cells of most microbial species. More severe heating is generally needed to inactivate or control spores of such microorganisms as the clostridia, including *C. botulinum*, and some bacilli (International Commission on Microbiological Specifications for Foods, 1980, pp. 1-37, 136-159).

Pasteurized cured products--e.g., frankfurters or canned hams--need refrigeration to avoid the possibility of their becoming toxic or spoiled from the development of surviving spores of *C. botulinum* or other organisms. Nitrite delays toxigenesis if such products are inadvertently temperature-abused. Products, such as deviled ham and Vienna sausage, that receive a botulinum (12D) cook are termed "commercially sterile," i.e., free from pathogens and organisms capable of growing under normal nonrefrigerated storage conditions. The so-called shelf-stable cured products receive a heat treatment that alone is not sufficient to inactivate all spores of *C. botulinum*, and the ability of these products to resist toxigenesis or spoilage depends on the presence of curing salts, such as sodium chloride and sodium nitrite (Pivnick *et al.*, 1969; Riemann, 1963). Heat treatment inactivates most cells and some spores, but those surviving are often injured or damaged and thus more susceptible to the inhibitory effects of sodium chloride or sodium nitrite (Roberts and Ingram, 1966; Roberts *et al.*, 1966). Heat injury will probably potentiate the effectiveness of other antimicrobial compounds or treatments that might serve as alternatives to nitrite, but this needs to be verified experimentally.

The possibility that the heating of an antimicrobial compound in a food (e.g., a cured meat) would lead to reactions that yield products with antimicrobial action needs to be considered. A strongly antibacterial "Perigo factor" is produced by the severe heating of nitrite in some laboratory media (Perigo *et al.*, 1967). It has been suggested that "Perigo-type" factors may be produced in thermally processed meat (Pivnick and Chang, 1973), but the evidence is inconclusive (National Academy of Sciences, 1981; Sofos *et al.*, 1979). This possibility needs to be investigated for compounds suggested as alternatives to nitrite.

Without the aid of other antimicrobial influences, properly applied severe thermal processing (at least 12D) can yield products that are not likely to spoil or become toxic. The large amount of commercially canned, low-acid vegetables--e.g., mushrooms, beans, and peas--safely consumed each year is testimony to that. Failure to deliver the intended heat treatment because of human error or mechanical failure is the most important threat to the safety of such products, and occasional product recalls have proved to be necessary and expensive. Promotion of the Hazard Analysis and Critical Control Point Program has greatly improved the reliability of thermal processing (Genigeorgis and Riemann, 1979).

In the case of cured-meat products that receive a botulinum cook, nitrite imparts color, antioxidant effects, possibly flavor, and, if processing is accidentally inadequate, some antimicrobial protection. However, most shelf-stable and pasteurized cured-meat products would be aesthetically unacceptable if subjected to a botulinum cook (Lechowich *et al.*, 1978). Sodium chloride and sodium nitrite are largely responsible for the safety of such products and their freedom from spoilage (Pivnick *et al.*, 1969).

MANIPULATION OF THE PRODUCT'S CHEMICAL OR PHYSICAL CHARACTERISTICS

The rate or extent of microbial proliferation in a cured product can be controlled not only by its components, which provide the substrates for growth, and by any added or naturally occurring antimicrobial agents but also by the chemical and physical characteristics of the product, i.e., its pH, water activity, and redox potential. These characteristics vary among products, and the microbial species and strains of concern in cured products vary in the extent to which they may be influenced by these characteristics (Chapter 3; International Commission on Microbiological Specifications for Foods, 1980, pp. 70-125, 136-159).

In the majority of products to which nitrite is added, no factor would be sufficiently inhibitory to eliminate proliferation of all or even most undesired microorganisms, unless that factor were drastically altered from the values associated with current production processes (Chapter 3). However, acting in concert, the various inhibitory influences (i.e., low temperature, heat injury, such preservatives as sodium chloride and sodium nitrite, pH, and water activity) combine to eliminate or greatly reduce microbial proliferation. A number of interactions are known to occur between such factors; they may be additive or truly synergistic or in some cases antagonistic, but they have not yet been sufficiently characterized. The nature of interactions is important in determining the potential usefulness of manipulating product characteristics as an alternative to adding nitrite to control microbial proliferation. Therefore, known interactions are mentioned in the following discussion, and some recommendations for future work in this connection have been included at the end of Chapter 3.

Acidity

Each microorganism has a characteristic optimal pH for growth above and below which growth decreases until values that limit growth are reached. Only a few foods are alkaline, and use of pH high enough to control microorganisms is not relevant to a discussion of meat products. However, the acidity of foods has for centuries been increased (and the pH lowered) naturally by fermentation or deliberately by the addition of acids. In some products, including a few cured meats, acidity may be the primary factor in preservation, but usually a suboptimal pH is complementary to other inhibitory factors (International Commission on Microbiological Specifications for Foods, 1980, pp. 92-111, 136-159, 333-409). Throughout the following discussion, it should be borne in mind that the growth of some organisms, such as molds, may result in microenvironments of altered (higher) pH in some products (e.g., acid tomato products). These microenvironments may be more favorable to proliferation of pathogens than the general product pH. This possibility exists for cured meats.

The pathogen of key concern in most cured products and many other preserved (e.g., canned) foods is *C. botulinum*. Foods are widely classified as low-acid or high-acid, with a pH of 4.6 as the dividing line. Below this line, outgrowth of *C. botulinum* spores and cell multiplication will not occur in most media (Odlaug and Pflug, 1970). Raatjes and Smelt (1979) recently reported growth in a particulate optimal complex medium whose pH was 4.2, but that is probably of little significance with respect to the safety or spoilage of cured products, inasmuch as all contain other inhibitory factors that act in concert to raise the minimal pH needed for inhibition of spore outgrowth and of cell multiplication. Decreasing pH below 7.0 increases the inhibitory effects of salt on vegetative cells of *C. botulinum* (Baird-Parker and Freame, 1967) and the effects of nitrite on a variety of microorganisms (Castellani and Niven, 1955; Tarr, 1941a,b, 1942, 1944).

The pH of most cured meats is 5.5-6.6. On rare occasions, it is as high as about 7.0-7.2 (International Commission on Microbiological Specifications for Foods, 1980, pp. 333-409). The role of pH in the inhibition of microbial proliferation and the way in which the lowering of pH is achieved vary among cured products. Some products, such as traditionally produced fermented sausages (e.g., salami and summer sausage), are often made at ambient temperatures. Lactic-acid-producing organisms that are naturally present as contaminants in the sausage mix are relied on to ferment carbohydrates in the formulation to lactic acid. In other products, such as bacons made with added carbohydrate (e.g., 0.5%), acid production by these bacterial contaminants occurs rapidly only when the product is removed from refrigeration and subjected to temperature abuse. Some organisms may be capable of slow growth below 10°C and ultimately reduce product pH under refrigeration.

In both these cases, the extent of acid production and the eventual pH depend on the occurrence of contamination of the product with lactic-acid-producing organisms and the amount of fermentable carbohydrate occurring naturally in or introduced into the product. In the case of the perishable products, in which naturally occurring contamination elicits a decrease in pH during temperature abuse (or slowly under refrigeration), the protection provided against pathogen growth is fortuitous. But attempts (described below) to use biologic acidulation as an alternative to conventional nitrite addition appear promising. Rapid establishment of fermentation (acidulation) in fermented sausage formulations has long been recognized as promoting product safety. Greater control over the biologic acidulation in fermented sausage is obtained by the use of inocula derived from previous successful fermentations or by the use of "starter cultures" that consist of laboratory cultures of the organisms desired in the fermentation (U.S. Department of Agriculture, 1977).

In lieu of reliance on biologic acidulation by naturally occurring or added lactic-acid-producing organisms, chemical acidulation is sometimes used in fermented sausages. As far as the Committee could determine, there are no products other than fermented sausages and some special products (such as souse) in which chemical acidulation is currently used commercially on a large scale.

Reliance on Naturally Occurring Acidulation. The major category of product in which the lowering of pH provides protection against pathogens and spoilage is the group of dried and semidried fermented sausages, such as salami and summer sausage (Hauschild, 1980). The various processes used in their production have been described by the USDA Task Force on the Staphylococci Enterotoxin Problem (U.S. Department of Agriculture, 1977). The Task Force also categorized the risk associated with those processes. In fermented sausages, the major pathogenic hazard is related to *Staphylococcus aureus*, the decrease in pH and microbial competition generally being sufficient to eliminate outgrowth and multiplication of *C. botulinum*. The hazard of staphylococci enterotoxin production arises during undesirable staphylococcal growth in the outer (aerobic) layers of the product. Conditions favorable to this growth include a high degree of contamination and production temperatures that favor the growth of *S. aureus* (e.g., above 26.4°C), rather than the growth of the naturally occurring lactic-acid-producing bacteria, which grow well at temperatures of 21-24°C. Most critical, however, appears to be failure to establish rapidly an active lactic-acid-producing fermentation or to attain a pH of 5.3 or less within 48 h of the initiation of fermentation.

The Task Force classified processes for producing fermented sausage as having high, intermediate, or low risk, with respect to staphylococcal enterotoxin hazard. Of five general types of production processes for fermented sausages, three rely on naturally occurring contamination. The traditional long process was classified as having low risk, because it used temperatures of 21°C or lower, which are unfavorable to staphylococci growth; the other two processes used higher incubation temperatures (26°C or higher) to shorten production times and were classified as having high risk. The Task Force recommended the use of chemical acidulation or starter cultures in such products as a means of reducing pH more rapidly and reducing the staphylococci health hazard. These same recommendations would be appropriate for control of _C. botulinum_ in similar products.

In a recent USDA four-plant study, the potential value of pH reduction by naturally occurring contamination as a protection against _C. botulinum_ was demonstrated. In this study, one set of bacon samples, inoculated with _C. botulinum_ spores, contained 0.57% added sucrose; other sets contained lower concentrations of added sugars or none at all. Samples were subjected to simulated temperature abuse by incubation at 27°C. Only 0.5% of the samples containing 0.57% sucrose became toxic during the 56-d incubation, whereas 30-95% of the samples containing less sugar became toxic. Within the first 28 d of the incubation, the pH of the sucrose-containing bacon was reduced to 5.0 or lower. The pH of the bacon containing the lower concentrations of sugar remained high (5.4 or above) throughout the entire 56 d (U.S. Department of Agriculture, 1979).

The foregoing discussion illustrates that naturally occurring contamination by lactic-acid-producing bacteria can contribute to lowering the pH of products that contain adequate fermentable carbohydrate, thereby increasing their resistance to pathogenic (staphylococcal and botulinal) hazard. The presence of a minimal concentration of fermentable carbohydrate is important for this acidulation. The reduction in pH needed to ensure freedom from microbial hazard depends on the other inhibitory factors in the product. Too little is known about the interactions of these inhibitory factors to predict the concentration of fermentable carbohydrate that would be ideal in this regard. However, 0.5-0.75% appears to be satisfactory in many cases. The degree of natural contamination also influences the speed and extent of pH reduction. Increased hygiene during production decreases contamination and thus reduces the likelihood of this fortuitous protective mechanism in products like bacon.

Controlled Biologic Acidulation. The acidulation that occurs when lactic-acid-producing bacteria ferment carbohydrates in a product is much more reliably managed when the responsible organisms and the substrate are both controlled. In the last few decades, producers of fermented sausages have adopted the use of a variety of "starter cultures" to facilitate control and reliability of production

(International Commission on Microbiological Specifications for Foods, 1980, pp. 136-159, 333-409; U.S. Department of Agriculture, 1977). Bacus and Brown (1981) and Coretti (1977) have recently reviewed the use of such cultures. The cultures can eliminate or greatly reduce the potential hazard from fermented sausage associated with *C. botulinum* spores and *S. aureus* (Christiansen *et al.*, 1975; U.S. Department of Agriculture, 1977).

A number of processes have been designed to use controlled biological acidulation in perishable products for protection against the proliferation of pathogens during temperature abuse. Riemann *et al.* (1972) described a method that used *Pediococcus cerevisiae* cells inactivated by irradiation. If added to the product with glucose, enzymes from these cells produced acid in the product during temperature abuse, but not under refrigeration. The resulting decrease in pH controlled pathogen growth.

Christiansen *et al.* (1975) examined the effect of sodium nitrite, starter culture, and glucose addition on *C. botulinum* growth and toxin production in a summer sausage. Samples formulated without glucose had a pH of 5.7-5.8 a week after storage at 27°C; those formulated with glucose had a pH of 4.4-5.1. Only 10 of 50 samples formulated without nitrite but with glucose became toxic in the total 112-d storage period at 27°C. Addition of sodium nitrite at 50 mg/kg or higher prevented toxin production (none of 125 samples were toxic) throughout the 112 d of storage. Of samples formulated without glucose with or without nitrite, most (115 of 175) became toxic. Increasing the concentration of nitrite in glucose-free products to 150 mg/kg reduced the proportion of samples that became toxic, but did not totally prevent toxin development. Table 5-1 shows these results. The importance of fermentable carbohydrate in such products and the benefits to be gained from using starter cultures can be seen. Sodium nitrite, even at relatively low concentrations, can decrease the likelihood of toxicity, especially if starter cultures are not used.

Tanaka *et al.* (1980) described the protection against formation of botulinum toxin in bacon afforded by the addition of lactobacilli, sucrose, and nitrite in various combinations; their results are shown in Table 5-2. Virtually no protection was afforded by sucrose alone, lactobacilli alone, or sodium nitrite alone at 40 mg/kg. Excellent protection--better than that with sodium nitrite at 120 mg with no added sucrose or lactobacilli--was offered by the combination of sucrose and lactobacilli with or without nitrite at 40 or 120 mg/kg. This method has the added advantage, over the use of killed cells, of increased effectiveness under temperature-abuse conditions because of the multiplication of lactic-acid-producing microorganisms that occurs at such temperatures.

A similar process was developed for bacon to lower product pH as a means of depleting nitrite more rapidly than occurs in conventional products; the increased depletion leads to a reduction in nitrosamine

TABLE 5-1

Botulinum Toxin Developed in Sausage Held at 27°C[a]

Nitrite Concentration, mg/kg of meat	Starter Culture	Glucose	No. Toxic/No. Tested At: (Days) 7	14	21	28	49	56	112	Total
0	X	X	0/3	0/3	0/3	0/5		1/5	1/6	2/25
50	X	X	0/3	0/3		0/5		0/5	0/9	0/25
150	X	X	0/3	0/3		0/5		0/5	0/7	0/23
0		X	1/3	0/3	2/3	3/5		0/5	2/6	8/25
50		X	0/3	0/3		0/5		0/5	0/9	0/25
100		X	0/3	0/3		0/5		0/5	0/9	0/25
150		X	0/3	0/3		0/5		0/5	0/8	0/24
0	X		0/3	22/22[b]						22/25
50	X		0/3	22/22[b]						22/25
100	X		0/3	0/3	0/4	2/5	2/10[b]			4/25
150	X		0/3	0/3	1/3	5/5	6/11[b]			15/25
0			0/3	1/3	19/19[b]					20/25
50			0/3	3/3	18/19[b]					21/25
150			0/3	0/3	2/2	3/5	9/12[b]			14/25

[a]From Christiansen *et al.*, 1975. Reprinted from Journal of Food Science, with permission. 1975. 40:488-490. Copyright © by the Institute of Food Technologists.
[b]All swollen.

TABLE 5-2

Effects of Lactobacilli, Sucrose, and Nitrite Addition to Bacon[a] Inoculated with Spores of *C. botulinum*[b] and Incubated at 27°C[c]

Sodium Nitrite, mg/kg	Sucrose[d]	Lactobacilli, Approximately 4×10^6 Cells/g[e]	Toxic Samples/ Samples Tested	%Toxic Samples	Time to Toxin
0	-	+	26/27	96	Most <1 wk
0	+	-	50/52	96	Most <1 wk
0	+	+	1/49	2.0	In 4th wk
40	-	-	47/50	94	Some <1 wk
40	+	+	0/30	0	
120	-	-	17/28	61	Some <1 wk Most <2 wk
120	-	+	34/68	50	Some <1 wk Most <2 wk
120	+	-	4/149	2.7	NA[f]
120	+	+	1/192	0.5	NA[f]

[a]Data from Tanaka *et al.* (1980) on commercially prepared bacon and that prepared specifically for the experiment, with permission. Data were aggregated.
[b]Approximately 1,000 spores/gram.
[c]Total incubation time, 8 wk.
[d]Sucrose concentrations were 0.5% or higher.
[e]*Lactobacillus plantarum*.
[f]NA = not available.

formation when the product is cooked (Bacus, 1979). A microbial culture is added to the bacon curing pickle; during smoking and storage, the pH of the product falls, and that leads to the desired reduction in the nitrite concentration of the final product and a lower concentration of nitrosamines after cooking, as shown in Table 5-3. This process has been approved for reducing nitrosamine concentrations in bacon (U.S. Department of Agriculture, 1979).

TABLE 5-3

Typical Analyses[a] of Bacon after 21 Days at 4.4°C

Formulation	Residual Nitrite, mg/kg	Standard Plate Count	pH	Nitrosopyrrolidine, μg/kg
Without starter culture	20-40	10^4-10^5	6.0-6.4	10-30
With starter	4-16	10^6-10^7	5.2-5.6	2-9

[a]Typical range of values observed; data from Bacus and Brown, 1981. Reprinted from Food Technology. 1981. Copyright c by the Institute of Food Technologists.

The possibility that biologic acidulation could be of use in providing protection against pathogens and spoilage of pickle-cured primal meat products is illustrated by the review of Coretti (1977) on the use of starter cultures in whole meats in Europe and the work of Bartholomew and Blumer (1977). The latter investigated the use of starter cultures to achieve more rapidly the sensory characteristics of dry-cured hams. However, a small number of hams containing starter culture were inadvertently subjected to temperature abuse during the investigation. They did not become spoiled, as would hams produced by the conventional process and similarly mistreated. A number of other reports that indicated the potential usefulness of biologic acidulation against spoilage and pathogens have been reviewed by Bacus and Brown (1981).

Chemical Acidulation. In certain fermented sausages, chemical acidulation is used to achieve an environment favorable to the growth of the desired flora (International Commission on Microbiological Specifications for Foods, 1980, pp. 136-159, 391; U.S. Department of Agriculture, 1977). In the United States, glucono-δ-lactone is the primary compound approved for this purpose. In some European countries, lactic acid is used as well.

Glucono-δ-lactone slowly hydrolyzes to gluconic acid when added to meat emulsions; at the permitted concentrations (5,000 mg/kg), it rapidly reduces pH by about 0.3-0.6 and establishes a pH of 5.2-5.4 (International Commission on Microbiological Specifications for Foods, 1980, p. 143; A. Petrica, Durkee Foods, Cleveland, Ohio, personal communication, 1981; Rugala, 1978). Chemical acidulation may be used in fermented sausage either alone or in conjunction with microbial fermentation. If careful attention is paid to formulation and distribution, chemical acidulation can effectively control the growth of _S. aureus_ and _C. botulinum_ (U.S. Department of Agriculture, 1977). Addition of lactic acid results in a greater reduction in pH, which may be desired in some cases.

In a very small number of products (e.g., pickled, cured products), acetic acid is used to lower pH sufficiently to inhibit the proliferation of pathogens and spoilage microorganisms. Information was presented to the Committee (Strumskis, 1981) on the use of a combination of acetic acid and glycerol to lower the pH and water activity of bacon to the extent that _C. botulinum_ growth is inhibited. The added acetic acid volatilizes during cooking.

Discussion. Manipulation of pH is a proven means of reducing spoilage and pathogenic hazards. It is currently used in a number of cured-meat products, in combination with nitrite addition. The relative contributions of pH and nitrite to control vary with the product and the pathogen of concern.

The use of acidulation of cured products as an alternative means of controlling undesired microbial proliferation could be extended by either of two means: additional chemical or biologic acidulation could be integrated into the production process so that the final product pH would be lower than that produced by conventional practices; or lactic-acid-producing organisms or systems (e.g., killed cells or enzymes) and carbohydrates could be incorporated into perishable products to cause acid to be produced under temperature abuse. The Committee believes that both these possibilities have potential utility as alternatives to the current use of nitrite. However, a number of considerations limit their immediate usefulness.

To replace added nitrite wholly or partly, products could theoretically be manufactured with an initial pH lower than that now characteristic. Some problems must, however, be noted. First, although decreasing pH in most products would probably increase the inhibition of pathogens shortly after processing, not enough is known about the interactions of the various antimicrobial influences (pH, nitrite, salt, a_w, etc.) to permit prediction in all products of the extent to which this would allow nitrite to be reduced. Second, decreasing pH would increase the rate of dissipation of nitrite in products. In consequence, residual nitrite during storage would fall more frequently than at present below the minimum necessary to inhibit remaining viable spores (Christiansen et al., 1978). Thus, protection against delayed temperature abuse might decrease with decreased product pH. Third, the extent to which the pH of a product can be reduced without loss of its characteristic sensory identity is not known. Finally, it should be noted that nitrosation reactions occur more rapidly at lower pH.

Water Activity

The water activity of a product can be thought of as a measure of the water available for biologic reactions, including microbial growth. Water activity (a_w) is defined as the ratio of the vapor pressure of water in the product (p) to that of pure water (p_o), i.e., $a_w = p/p_o$. Groups of organisms have ranges within which growth of most members of the group will occur (Table 5-4), and each species and strain has its own characteristic range. The limits of this range may be reduced if other environmental conditions, such as pH and temperature, are not optimal. The International Commission on Microbiological Specifications for Foods (1980, pp. 70-91) has reviewed the influence of water activity on pathogens and spoilage organisms of concern in foods, including cured meats.

TABLE 5-4

Approximate Lower Limits of Water Activity for Microbial Growth[a]

Microbial Group	Water Activity
Bacteria	0.86-1.0
Yeasts	0.73-0.88
Molds	0.61-0.78

[a]Modified from Troller, 1980.

A variety of factors affect water activity, including the moisture content of the product and the nature of solutes in the aqueous phase. The water activity of a product is most commonly manipulated through drying (as with dry or semidry sausages) or through the addition of such solutes as sodium chloride and sugar. A combination of these methods may contribute to the final product water activity. The water activity of most cured meats is higher than 0.95. However, in some products it may be lower, e.g., when the brine concentration exceeds 10%, which corresponds to a water activity of 0.92 (International Commission on Microbiological Specifications for Foods, 1980, pp. 333-409); such products include most dry or semidry sausages and some dry-cured cuts.

The strains of *C. botulinum* that produce various types of toxin differ in their resistance to lowered water activity. In media whose water activity was adjusted with sodium chloride, the lower limits of growth from spore inocula were 0.95a_w for type A, 0.94a_w for type B, and 0.97a_w for type E. Toxin production occurs at water activities approaching those which inhibit growth (Ohye and Christian, 1967). Some agents commonly used to reduce water activity, such as sodium chloride, may also have additional preservative action. Baird-Parker and Freame (1967) reported that growth of all three types noted above was inhibited at higher water activities when sodium chlorde was used, rather than glycerol. Acott *et al.* (1976) discussed other humectants with possible antimicrobial activity.

Staphylococcus aureus is the pathogen of greatest concern in cured products with low water activity, because it has been reported to have a lower limit in the range of 0.83-0.86 under aerobic conditions, depending on the medium. This is considerably lower than the water activity of most cured foods. Limits on enterotoxin production have been discussed by the International Commission on Microbiological Specifications for Foods (1980, p. 84). As discussed previously, the staphylococcal enterotoxin hazard in fermented sausages is amenable to control through the use of microbial competition or acidulation.

The use of water activity in conjunction with pH has been suggested as a basis for categorizing meat products (Leistner and Rödel, 1975). In some products, such as dry fermented sausages or dry-cured hams, the water activity may be sufficiently low that it alone eliminates the potential hazard from *C. botulinum* toxigenesis, or it may play a predominant role in this respect in concert with other inhibitory factors.

Similar considerations apply to the potential utility of reducing the water activity of products as a means of reducing the need for added nitrite as are pertinent to pH modification. Water-activity reduction in many products would probably allow a reduction in added nitrite without lessening antimicrobial protection. However, too little is known about the interactions of water activity with other inhibitory influences to permit predictions of the extent to which nitrite could be decreased. A lowering of water activity may also produce sensory changes in products, modifying their traditional texture and other attributes. Another problem that may restrict the usefulness of water-activity manipulation is the difficulty of achieving and maintaining uniform a_w in a food product. The non-uniformity of meat or meat products and the condensation that may occur on chilling contribute to the problem (International Commission on Microbiological Specifications for Foods, 1980, pp. 70-91).

Redox Potential

The control exerted over microbial proliferation by the redox potential of the medium has recently been reviewed by the International Commission on Microbiological Specifications for Foods (1980, pp. 112-125); however, very little is known about this influence on microbial proliferation in microbiologic media and especially in foods.

MAINTENANCE OF CONTROLLED ENVIRONMENTS

The multiplication of microorganisms may be eliminated by strictly maintaining an unfavorable environment. Low (or high) temperature and particular gaseous atmospheres can be used to this end. Maintenance of low temperature either by refrigeration or by freezing is the most practicable method in this category and is in wide use. The utility of controlling gaseous exposure through packaging is also discussed briefly below.

Low Temperature

The effects of low temperature on the proliferation of the various microorganisms of concern in foods, including cured meats, has recently been reviewed by the International Commission on Microbiological Specifications for Foods (1980, pp. 1-37). The minimal temperature at which growth occurs depends on the species of microorganism and often on the strains within a species. A wide variety of microorganisms are of potential concern as pathogens or spoilage organisms in cured meats, and nitrite contributes to the inhibition of some of them. Its contribution in the control of some of these varies with storage temperature. Rigorous and continuous application of the appropriate storage temperature can effectively control microbial growth or spore outgrowth.

Many raw or pasteurized cured products require refrigeration during distribution and storage. Storage at refrigeration temperatures has played an important role in the extension of shelf-life of some cured meats and in their excellent public-health record. The proteolytic strains of _C. botulinum_ that produce type A or B toxin are the primary health concern in cured meats. They do not multiply and produce toxin at temperatures below 10°C (Ohye and Scott, 1953). However, because products are often distributed through complex networks in the United States, one cannot be certain that a given product will not be temperature-abused during distribution or retailing or while in the consumer's possession--e.g., in the home, in packed lunches, or at picnics. It is evident that, if refrigeration were increasingly used as a means of reducing the addition of nitrite, the whole system would require strict controls, which would be difficult to monitor adequately.

Nonproteolytic strains of C. botulinum--e.g., type E, which occurs mostly in fish products--grow and produce toxin down to 3.3°C (Schmidt et al., 1962). Thus, if refrigeration is marginally inadequate, such strains pose a problem in fish and could do so in meats if contamination occurs. This, however, is offset somewhat by their greater sensitivity to salt (Genigeorgis and Riemann, 1979).

If, in the future, reliance were placed on improved refrigeration or freezing in combination with lower nitrite concentrations, such methods may lead to longer storage than is now used. However, ascorbate, which is added to bacon and some other products to promote color fixation and reduce nitrosamine formation, increases the depletion of nitrite during storage. When the sodium nitrite concentration is below about 30 mg/kg, C. botulinum spores that have germinated (as is possible during refrigeration) could produce toxin (Christiansen et al., 1978). Thus, a contaminated product kept refrigerated or frozen for a long period would pose a health hazard if it were subjected to temperature abuse. That is true with current nitrite addition, but would be exacerbated if nitrite addition were reduced and reliance were placed solely on refrigeration or freezing.

Unless refrigeration temperatures lower than those commonly used were adopted, refrigeration could not serve as a substitute for nitrite in the control of some psychrotrophic spoilage organisms (Terrell, 1974).

Like refrigeration, freezing (when closely monitored) might provide a good alternative to high concentrations of nitrite (Sofos and Busta, 1980). However, the freezing of heated or comminuted cured meats may result in the accelerated development of rancid flavor, owing to lipid oxidation (Kramlich et al., 1973). Because one of the roles of nitrite is to inhibit lipid oxidation, rancid cured meats might be more prevalent if low-nitrite products were frozen. However, MacDonald et al. (1980) demonstrated that lower concentrations of nitrite than those currently used did significantly reduce lipid oxidation in some cured meats. Today, some canned luncheon meats and hams with high concentrations of nitrite are often stored frozen for limited periods by the U.S. Army and consumers. Microbial growth does not occur below -10°C, but some enzymes remain active. These enzymes could limit the shelf-life. If the product is not subjected to temperature abuse, there should be no problem related to a microbial health hazard.

In addition to the qualifications noted above, in connection with the need to maintain low temperatures continuously for effective restriction of microbial proliferation in perishable products, one point needs emphasis. Products in one major class--shelf-stable canned meats, such as luncheon meat--are not refrigerated. Partial or complete omission of nitrite from these products, in favor of refrigeration to ensure safety, would probably change their sensory characteristics and necessitate extensive changes in distribution and consumer handling. Changes of this nature would have unpredictable safety consequences while consumers were adopting the new precautions.

In addition, residual nitrite in these products at the likely time of consumption (≥60 d) is lower than in other cured meats (National Academy of Sciences, 1981, Chapter 5).

Gaseous Environments

With rigorous control of the gaseous environment of a product, it is possible to eliminate or greatly reduce problems caused by some types of organisms (International Commission on Microbiological Specifications for Foods, 1980, pp. 170-192). For example, vacuum packaging will greatly reduce or delay spoilage caused by some aerobic mesophiles such as members of the Pseudomonas, Acinetobacter, and Moraxella genera (International Commission on Microbiological Specifications for Foods, 1980, pp. 333-409). Vacuum packaging aids in control of many spoilage organisms while the packaging seal remains intact, but it does not substantially increase the growth of anaerobic spoilage organisms or pathogens.

Maintenance of an environment in a packaged meat product that is sufficiently aerobic to inhibit the growth of obligately anaerobic organisms (mostly spore-formers) is not generally considered practicable in cured meats, for a number of important reasons. Cooked cured meats and especially sliced and comminuted products are susceptible to lipid oxidation, and an aerobic environment would promote rancidity more rapidly in such products (see Chapter 6). The reddish-pink color of cured meats would also be shorter-lived. With regard to microbial pathogens, it has been demonstrated that growth of *Staphylococcus aureus* and production of its enterotoxin are favored by aerobic environments (Crowther *et al.*, 1977). Most importantly, however, Christiansen and Foster (1965) demonstrated that spores of *C. botulinum* type A inoculated between slices of bologna germinated and produced toxin in aerobic incubation. An anaerobic microenvironment must have been generated in the product for this to occur. More recently, Kautter *et al.* (1981) demonstrated botulinal toxin production in sandwiches packed aerobically or under an atmosphere of nitrogen.

The use of gases as preservatives for foods, including meat products, has been reviewed by the International Commission on Microbiological Specifications for Foods (1980, pp. 170-192). Carbon dioxide at 20% prolongs the shelf-life of refrigerated fresh meats by inhibiting pseudomonads and the Acinetobacter-Moraxella group (Silliker and Wolfe, 1980). This may be useful for some cured meats. Sulfur dioxide displays useful antifungal and antibacterial activity in some products, including sausages, but in the U.S. is not permitted in foods recognized as sources of thiamine. The use of sulfur dioxide and related salts is discussed in Chapter 4. Carbon monoxide appears to have an inhibitory effect similar to that of carbon dioxide on psychrotrophic bacteria and to prolong the color of cured meats (Clark *et al.*, 1976; International Commission on Microbiological Specifications for Foods, 1980, pp. 170-192; Silliker and Wolfe, 1980).

CONCLUSIONS

Irradiation alone or in conjunction with low concentrations of nitrite can be used to produce microbiologically safe bacon, corned beef, and ham. It is also effective against staphylococci and salmonellae, which are poorly controlled by nitrite.

Severe thermal processing of products in the final container (i.e., ones not repackaged) can ensure their microbiologic safety; but most cooked cured items do not receive such a treatment, because it would be detrimental to their traditional aesthetic and other characteristics.

Reduction in product pH or water activity is used, and possibly could be used more widely, to increase protection against pathogens. Low pH or water activity is used routinely in some products, e.g., in fermented sausages, where these factors are the primary ones inhibiting microbial proliferation. Too little is known about the interaction of various factors controlling microbial proliferation in cured meats to predict the extent to which reducing pH or water activity of other products would enable the concentrations of nitrite to be reduced while maintaining antimicrobial protection. Such reductions might affect the sensory characteristics of products (Chapter 9).

Temperature-abuse-activated biologic acidulation (from fermentable carbohydrate) by lactic-acid-producing bacteria has been demonstrated in commercially prepared bacon to be as effective as conventional nitrite use. This method is equally or more effective when used in combination with sodium nitrite at 40 mg/kg and appears applicable to other products.

Both freezing and refrigeration need to be maintained continuously to provide protection against microbial proliferation. This cannot be guaranteed. Thus, they cannot be considered as total replacements for nitrite, which provides protection during temperature abuse. Partial or complete omission of nitrite from shelf-stable canned cured meats in favor of refrigeration would probably change their sensory characteristics. Introducing the necessity for refrigerating such products would have unpredictable safety consequences during the period when producers, consumers, and distributors are adopting the new precautions.

Although manipulations of the redox potential of products or of the gaseous atmosphere in which they are packaged may increase the shelf-life of cured products, they cannot yet be judged to be feasible alternatives to nitrite in the control of the major problems caused by anaerobic spoilage organisms or pathogens in cured meats.

The Committee recognizes that it has not addressed (or has addressed only briefly) a number of theoretically possible complete or partial alternatives to the current use of nitrite. These include changes in production, such as requiring thermal processing of all cured products sufficient to ensure commercial sterility. The Committee excluded this possibility from its deliberations, because it would substantially alter the aesthetic qualities of the vast majority of cured products. In addition, it was unable, because of time constraints, to consider other possibilities fully, such as the use of indicators on packages to warn the consumer that a product had been subjected to temperature abuse. If consumers could be relied on to heed them, such indicators might be helpful while the product remained in the original package. However, consumer compliance with suggested practices cannot be guaranteed.

RECOMMENDATIONS

Investigation of the use of irradiation in products other than bacon, corned beef, and ham is needed, as is study of its application to commercial production.

High priority should be accorded to investigations of the interaction of various factors that contribute to the control of microorganisms in products to which nitrite is added. Heating, pH, and water activity should be among the factors included in such investigations, details of which are suggested in Chapter 3.

The Committee endorses a previous recommendation (National Academy of Sciences, 1975, p. 84) that producers of fermented sausages should be encouraged to use chemical acidulation or starter cultures where appropriate to promote product safety. Consideration should be given to approval of additional acidulants for use in meat products.

The efficacy of temperature-abuse-activated biologic acidulation (from fermentable carbohydrate) by lactic-acid-producing microorganisms for antibotulinal protection in bacon should be investigated with a variety of microorganisms under several commercial production conditions. The applicability of this method to other products should also be evaluated.

The Committee recommends further investigations of various known humectants (e.g., glycerol, glycols, sodium chloride, 1,3-butadiene, and polyhydric alcohols) to determine their suitability as agents for modifying the water activity of cured meats. The antimicrobial efficacy of the various means of reducing water activity should also be studied, because there is evidence of variation in antimicrobial effectiveness among humectants. Some of these possibilities, such as salt, have known toxicologic liabilities, and all would require careful evaluation before any increase in human exposure could be justified.

Because correct and continuous refrigeration of perishable cured products could help to eliminate the risk of botulism from those products and thus help to reduce the need for nitrite's antibotulinal protection, the Committee recommends increased efforts by regulatory agencies and producers to educate consumers and producers with respect to proper refrigeration, possibly through more prominent labeling.

Investigations of the use of gases, such as nitrogen, or packaging under a vacuum to increase the shelf-life of cured products should be encouraged.

REFERENCES

Acott, K. M., A. E. Sloan, and P. T. Labuza. 1976. Evaluation of antimicrobial agents in a microbial challenge. Study for an intermediate moisture dog food. J. Food Sci. 41:541-546.

Anellis, A., D. Berkowitz, W. Swantak, and C. Strojan. 1972. Radiation sterilization of prototype military foods: Low-temperature irradiation of codfish cake, corned beef and pork sausage. Appl. Microbiol. 24:453-462.

Anellis, A., E. Shattuck, T. Latt, S. Songpasertchal, D. B. Rowley, and E. W. Ross, Jr. 1977. Gamma irradiation at $-30 \pm 10^{\circ}$ of low level nitrite/nitrate ham. Pp. 631-647 in A. N. Barker, J. Wolf, D. J. Ellar, G. J. Dring, and G. W. Gould, eds. Spore Research 1976, Volume II. Academic Press, London.

Anellis, A., D. B. Rowley, and E. W. Ross, Jr. 1979. Microbiological safety of radappertized beef. J. Food Prot. 42-927-932.

Bacus, J. N. 1979. Reduces nitrosamines. Food Eng. 51:24.

Bacus, J. N., and W. L. Brown. 1981. Use of microbial cultures: Meat products. Food Technol. 36:74-78, 83.

Baird-Parker, A. C., and B. Freame. 1967. Combined effect of water activity, pH, and temperature on the growth of *Clostridium botulinum* from spore and vegetative cell inocula. J. Appl. Bacteriol. 30:420-429.

Bartholomew, D. R., and T. N. Blumer. 1977. The use of a commercial *Pediococcus cerevisiae* culture in the production of country-style hams. J. Food Sci. 42:494-497.

Castellani, A. G., and C. F. Niven, Jr. 1955. Factors affecting the bacteriostatic action of sodium nitrite. Appl. Microbiol. 3:154-159.

Christiansen, L. N., and E. M. Foster. 1965. Effect of vacuum packaging on growth of *Clostridium botulinum* and *Staphylococcus aureus* in cured meats. J. Appl. Microbiol. 13:1023-1025.

Christiansen, L. N., R. B. Tompkin, A. B. Shaparis, R. W. Johnston, and D. A. Kautter. 1975. Effect of sodium nitrite and nitrate on *Clostridium botulinum* growth and toxin production in a summer style sausage. J. Food Sci. 40:488-490.

Christiansen, L. N., R. B. Tompkin, and A. B. Shaparis. 1978. Fate of *Clostridium botulinum* in perishable canned cured meat at abuse temperature. J. Food Prot. 41:354-355.

Clark, D. S., C. P. Lentz, and L. A. Roth. 1976. Use of carbon monoxide for extending shelf life of prepackaged fresh beef. Can. Inst. Food Sci. Technol. J. 9:114-117.

Coretti, K. 1977. Starterkulturen in der Fleischwirtschaft. Die Fleischwirtschaft 3:386-394.

Crowther, J. S., R. Holbrook, A. C. Baird-Parker, and B. L. Austin. 1977. Role of nitrite and ascorbate in the microbiological safety of vacuum-packed sliced bacon. Pp. 13-20 in B. J. Tinbergen and B. Krol, eds. Proceedings of the 2nd International Symposium on Nitrite in Meat Products, Zeist, PUDOC, Wageningen, the Netherlands.

Genigeorgis, C., and H. Riemann. 1979. Food processing and hygiene. Pp. 613-713 in H. Riemann and F. L. Bryan, eds. Foodborne Infections and Intoxications, 2nd Ed. Academic Press, New York. 748 pp.

Hauschild, A. H. W. 1980. Microbiological problems in food safety with particular reference to _Clostridium botulinum_. Pp. 68-107 in H. D. Graham, ed. The Safety of Foods, 2nd Ed. AVI Publishing, Westport, Connecticut.

Ingram, M., and J. Farkas. 1977. Microbiology of foods pasteurized by ionizing radiation. Acta Aliment. 6:123-185.

International Commission on Microbiological Specifications for Foods. 1980. Microbial Ecology of Foods, Volumes 1 (pp. 1-332) and 2 (pp. 333-998). Academic Press, New York.

Kautter, D. A., R. K. Lynt, and T. Lilly. 1981. Evaluation of botulism hazard from nitrogen-packed sandwiches. J. Food Prot. 44:59-61.

Kramlich, W. E., A. M. Pearson, and F. W. Tauber. 1973. Processed Meats. AVI Publishing Co., Westport, Connecticut.

Lechowich, R. V., W. L. Brown, R. H. Deibel, and I. I. Somers. 1978. The role of nitrite in the production of canned cured meat products. Food Technol. 32:45-58.

Leistner, L., and W. Rödel. 1975. The significance of water activity for micro-organisms in meats. Pp. 309-323 in R. B. Duckworth, ed. Water Relations of Foods. Academic Press, New York, San Francisco.

MacDonald, B., J. I. Gray, D. W. Stanley, and W. R. Usborne. 1980. Role of nitrite in cured meat flavor: Sensory analysis. J. Food Sci. 45:885-888, 904.

National Academy of Sciences. 1975. Prevention of Microbial and Parasitic Hazards Associated with Processed Foods: A Guide for the Food Processor. National Academy of Sciences, Washington, D.C. 166 pp.

National Academy of Sciences. 1981. The Health Effects of Nitrate, Nitrite, and N-Nitroso Compounds. National Academy Press, Washington, D.C. 544 pp.

Odlaug, T. E., and I. J. Pflug. 1979. *Clostridium botulinum* and acid foods. J. Food Prot. 41:566-573.

Ohye, D. F., and J. H. B. Christian. 1967. Combined effects of temperature, pH and water activity on growth and toxin production by *C. botulinum* types A, B and E. Pp. 217-223 in M. Ingram and T. A. Roberts, eds. Botulism 1966. Chapman and Hall, London.

Ohye, D. F., and W. J. Scott. 1953. The temperature relations of *Clostridium botulinum*, types A and B. Aust. J. Biol. Sci. 6:178-189.

Perigo, J. A., E. A. Whiting, and T. E. Bashford. 1967. Observations on the inhibition of vegetative cells of *Clostridium sporogenes* by nitrite which has been autoclaved in a laboratory medium discussed in the context of sublethally processed cured meats. J. Food Technol. 2:377-397.

Pivnick, H., and P.-C. Chang. 1973. Perigo effect in pork. Pp. 111-116 in B. Krol and B. J. Tinbergen, eds. Proceedings of the International Symposium on Nitrite in Meat Products, Zeist, PUDOC, Wageningen, the Netherlands.

Pivnick, H., H. W. Barnett, H. R. Nordin, and L. J. Rubin. 1969. Factors affecting the safety of canned, cured, shelf-stable luncheon meat inoculated with *Clostridium botulinum*. Can. Inst. Food Sci. Technol. J. 2:141-148.

Raatjes, G. J. M., and J. P. P. M. Smelt. 1979. *Clostridium botulinum* can grow and form toxin at pH values lower than 4.6. Nature 281:398-399.

Riemann, H. 1963. Safe heat processing of canned cured meats with regard to bacterial spores. Food Technol. 17:39-49.

Riemann, H., W. H. Lee, and C. Genigeorgis. 1972. Control of *Clostridium botulinum* and *Staphylococcus aureus* in semi-preserved meat products. J. Milk Food Technol. 35:515-523.

Roberts, T. A., and M. Ingram. 1966. The effect of sodium chloride, potassium nitrate and sodium nitrite on the recovery of heated bacterial spores. J. Food Technol. 1:147-163.

Roberts, T. A., R. J. Gilbert, and M. Ingram. 1966. The effect of sodium chloride on heat resistance and recovery of heated spores of Clostridium sporogenes (P.A. 3679/S_2). J. Appl. Bacteriol. 29:549-555.

Rowley, D. B., and A. Brynjolfsson. 1980. Potential uses of irradiation in the processing of food. Food Technol. 34(10): 75-77.

Rowley, D. B., R. Sullivan, and E. S. Josephson. 1978. Indicators of viruses in foods preserved by ionizing radiation. P. 355 in G. Berg, ed. Indicators of Viruses in Water and Food. Ann Arbor Science Publishers, Inc., Ann Arbor, Michigan.

Rowley, D. B., R. Firstenberg-Eden, E. M. Powers, G. E. Shattuck, A. E. Wasserman, and E. Wierbicki. In press. Irradiation: An alternative to high levels of nitrite in bacon. J. Food Sci.

Rugala, H. W. 1978. Encapsulation controls GLD release--production efficiency improved. Food Prod. Dev. 12(10):30-31.

Schmidt, C. F., W. K. Nank, and R. V. Lechowich. 1962. Radiation sterilization of food: II. Some aspects of the growth, sporulation, and radiation resistance of spores of Clostridium botulinum, type E. J. Food Sci. 27:77-84.

Silliker, J. H., and S. K. Wolfe. 1980. Microbiological safety considerations in controlled atmosphere storage of meats. Food Technol. 34(3):59-63.

Sofos, J. N., and F. F. Busta. 1980. Alternatives to the use of nitrite as an antibotulinal agent. Food Technol. 34(6):244-251.

Sofos, J. N., F. F. Busta, and C. E. Allen. 1979. Botulism control by nitrite and sorbate in cured meats: A review. J. Food Prot. 42:739-790.

Strumskis, L. 1981. Bacon Cured Using the CH...OH Radical Group of Acetic Acid and Glycerine Dehydrants. Paper presented at the Public Meeting of the National Academy of Sciences, Committee on Nitrite and Alternative Curing Agents in Food, January 22, 1981, Washington, D.C. 3 pp.

Stumbo, C. R. 1973. Thermobacteriology in Food Processing, 2nd Ed. Academic Press, New York. 329 pp.

Tanaka, N., E. Traisman, M. H. Lee, R. G. Cassens, and E. M. Foster. 1980. Inhibition of botulinum toxin formation in bacon by acid development. J. Food Prot. 43:450-457.

Tarr, H. L. A. 1941a. The action of nitrites on bacteria. J. Fish. Res. Board Can. 5:265-275.

Tarr, H. L. A. 1941b. Bacteriostatic action of nitrites. Nature (London) 147:417-418.

Tarr, H. L. A. 1942. Action of nitrates and nitrites on bacteria. J. Fish. Res. Board Can. 6:74-89.

Tarr, H. L. A. 1944. Action of nitrates and nitrites on bacteria. J. Fish. Res. Board Can. 6:233-242.

Terrell, R. N. 1974. Theory and practice of preblending for use in comminuted meat food mixes. Pp. 23-33 in Proceedings of Meat Industry Research Conference. American Meat Institute, Washington, D.C.

Troller, J. A. 1980. Influence of water activity on microorganisms in foods. Food Technol. 34:(5)76-80, 82.

U.S. Department of Agriculture. 1977. The Staphylococcal Enterotoxin Problem in Fermented Sausage. Food Safety and Quality Service, U.S. Department of Agriculture, Washington, D.C. 13 pp. plus appendixes.

U.S. Department of Agriculture. 1979. Acid producing microorganisms in meat products for nitrite dissipation. Fed. Reg. 44:9372.

CHAPTER 6

INHIBITION OF LIPID OXIDATION IN CURED MEAT

CHAPTER 6

INHIBITION OF LIPID OXIDATION IN CURED MEAT

Nitrite inhibits lipid oxidation in cured meats. Such an effect is desirable, because lipid oxidation may affect flavor, and the products of such oxidation may pose hazards to human health. In this chapter, the adverse effects of lipid oxidation are discussed briefly, methods commonly used to assess the extent of lipid oxidation in meat are discussed, and the antioxidant activities of nitrite and potential alternatives to nitrite are reviewed.

EFFECTS OF LIPID OXIDATION

Effects on Flavor

The fat (lipid) component of meat contributes substantially to its flavor and palatability. The basic meaty flavor resides in the water-soluble fraction of meat, but the flavor that distinguishes pork from beef or fish, for example, resides in the lipid fraction (Hornstein, 1967; Hornstein *et al.*, 1960). Lipids are easily oxidized, and such changes can markedly affect flavor; for example, the rancidity that develops in stored meat is due to lipid oxidation. Refrigerated, cooked meat can develop a rancid or stale flavor termed "warmed-over flavor" (WOF) (Tims and Watts, 1958) within 48 h at 4°C. This is in marked contrast with the slow onset of rancidity commonly encountered in raw meats, fatty tissues, rendered fat, or lard, which is normally not apparent until after storage for weeks or months (Pearson *et al.*, 1977). Although WOF has generally been recognized as pertaining only to cooked meat, there is now evidence that it develops just as rapidly in raw meat that has been ground and exposed to air (Sato and Hegarty, 1971).

Possible Adverse Effects on Health

Malonaldehyde (malondialdehyde) is formed by the oxidation of polyunsaturated fatty acids. Interest in this particular product of lipid oxidation has stemmed from reports that it reacts with DNA (Brooks and Klamerth, 1968), is mutagenic in the Salmonella/microsome assay (Mukai and Goldstein, 1976), and induces tumors in mice when it is dissolved in acetone and applied topically with croton oil as a promoter (Shamberger *et al.*, 1974). However, a recent report demonstrated that the mutagenic activity of earlier malonaldehyde preparations was due almost entirely, if not exclusively, to contaminants (Marnett and Tuttle, 1980). Enzymatically synthesized malonaldehyde (Summerfield and Tappel, 1978) is without mutagenic activity in the standard Salmonella/microsome assay tester strains (Ames *et* al., in press), although it does have weak mutagenic activity in a new Salmonella tester strain that reverts by frameshift mutation in a repeating sequence of adenine residues (D. Levin and B. N. Ames, University of Calfornia, Berkeley, personal communication). In addition, Apaja (1980) failed to demonstrate carcinogenicity of malonaldehyde administered to random-bred Swiss mice in drinking water at concentrations of 0.5, 0.25, and 0.125%. However, because of the high mortality in the high-dose group, there is still "a possibility for increased occurrence of late-developing tumors" at the higher dose.

Thus, the significance of the earlier reports on possible adverse effects of malonaldehyde is unclear. The importance of malonaldehyde formed by the oxidation of lipids in meats is even less clear, because most of it is bound to other molecules, and its fate after ingestion is unknown.

It is known, however, that most of the malonaldehyde ingested is derived from meat (Shamberger et al., 1977), and cured meats may be the greatest contributor, inasmuch as bacon, for example, contains about twice as much unsaturated fatty acid as saturated fatty acid (Anonymous, 1981) and about 19% of the unsaturated fatty acid in pork is polyunsaturated (Benedict, 1980).

The average daily intake of malonaldehyde from meats (cured and uncured) can be estimated at approximately 230 μg/person, on the basis of malonaldehyde concentrations reported by Siu and Draper (1978) and consumption figures developed by the Committee (National Academy of Sciences, 1981). This estimate may be low: Shamberger et al. (1977) reported malonaldehyde concentrations that were several times higher than those of Siu and Draper.

In summary, the significance for human health of malonaldehyde in meats is unknown, but the meager data available emphasize the desirability of minimizing the occurrence of malonaldehyde during storage and marketing. However, inhibition of malonaldehyde formation may pose health hazards as well. Nitrite, for example, decreases the concentrations of malonaldehyde in meat; however, in the presence of nitrite, malonaldehyde can facilitate nitrosation of secondary amines (Kikugawa et al., 1980; Kurechi et al., 1980). Nitrosation apparently occurs, in this case, through intermediate compounds--aminoacroleins--that are more readily nitrosated under mildly acidic conditions than are their parent amines (Kikugawa et al., 1980).

LIPID COMPOSITION OF MEATS

Meat lipids are commonly classified as adipose lipids or intramuscular (tissue) lipids (Pearson et al., 1977; Watts, 1962). The adipose lipids consist mainly of triglycerides; intramuscular lipids are composed of both triglycerides and membrane-bound lipids, such as phospholipids and lipoproteins (Love and Pearson, 1971; Pearson et al., 1977). Although the composition and structure of the adipose lipids are influenced by species, diet, environment, and sex (Deuel, 1965; Hilditch and Williams, 1964; Link *et al.*, 1970a,b), the composition and structure of intramuscular lipid generally do not vary, and only the amount present varies with the type of meat (Koch *et al.*, 1968; Orme *et al.*, 1958).

Phospholipids, present mainly in the intramuscular lipids, appear to be the components most susceptible to oxidation and are at least partially responsible for the off-flavors that develop during the storage of cooked meat (Younathan and Watts, 1959, 1960). Phospholipid concentrations of red and white muscle from various sources have been reported by Wilson *et al.* (1976) and are given in Table 6-1. On the basis of this study, phospholipid content is highest in the dark meat of chicken and white meat of pork. Apparently, the tendency for phospholipids to undergo rapid oxidation is due largely to their high content of unsaturated fatty acids (Giam and Dugan, 1965; O'Keefe *et al.*, 1968). Table 6-2 shows that the triglyceride fraction of meat is, in fact, lower in unsaturated fatty acids than is the phospholipid fraction (Hornstein *et al.*, 1961).

CHEMISTRY OF LIPID OXIDATION

The oxidation of unsaturated fatty acids--such as oleic, linoleic, and linolenic acids--generally proceeds through a free-radical chain mechanism involving initiation, propagation, and termination:

Initiation:

$$2LH + O_2 \longrightarrow 2L^{\bullet} + 2OH^{\bullet}$$

Propagation:

$$L^{\bullet} + O_2 \longrightarrow LOO^{\bullet}$$
$$LOO^{\bullet} + LH \longrightarrow LOOH + L^{\bullet}$$

Termination:

$$L^{\bullet} + L^{\bullet} \longrightarrow LL$$
$$L^{\bullet} + LOO^{\bullet} \longrightarrow LOOL$$
$$LOO^{\bullet} + LOO^{\bullet} \longrightarrow LOOL + O_2$$

LH refers to any unsaturated fatty acid; $L^{\bullet}$, to a free radical formed by removing a labile hydrogen from a carbon atom adjacent to a double bond; and LOOH, to a hydroperoxide, one of the major initial oxidation products that decompose to form products responsible for off-flavors (Gaddis *et al.*, 1961; Horvat *et al.*, 1969). Such products include hexanal, pentanal, and malonaldehyde, all of which have been detected in cooked, uncured meat (Cross and Ziegler, 1965; Tarladgis *et al.*, 1960).

TABLE 6-1

Mean Lipid Contents and Standard Deviations for Red and White Muscle from Mutton, Beef, Pork, Chicken, and Turkey[a]

Species	Total Lipid, % of Tissue Weight		Phospholipid, % of Total Lipid		Phospholipid, % of Tissue Weight	
	Red Muscle[b]	White Muscle[c]	Red Muscle[b]	White Muscle[c]	Red Muscle[b]	White Muscle[c]
Mutton	5.58 $\pm$ 0.49	--	17.25 $\pm$ 1.81	--	0.80 $\pm$ 0.14	--
Beef	14.79 $\pm$ 0.39	--	3.56 $\pm$ 0.16	--	0.50 $\pm$ 0.02	--
Pork	5.47 $\pm$ 0.62	8.88 $\pm$ 0.67	16.73 $\pm$ 3.35	11.97 $\pm$ 2.77	0.83 $\pm$ 0.01	1.00 $\pm$ 0.19
Chicken	4.74 $\pm$ 0.37	1.52 $\pm$ 0.15	42.25 $\pm$ 7.60	42.40 $\pm$ 4.70	1.60 $\pm$ 0.05	0.50 $\pm$ 0.03
Turkey	1.86 $\pm$ 0.15	0.79 $\pm$ 0.05	35.43 $\pm$ 3.11	64.42 $\pm$ 3.67	0.63 $\pm$ 0.02	0.55 $\pm$ 0.01

[a]Data from Wilson *et al.*, 1976. Reprinted with permission from Journal of Agriculture and Food Chemistry. Copyright © 1976. American Chemical Society. Mean values are from three animals of the same species.

[b]Red muscle as follows: mutton and beef, longissimus; pork, red portion of semitendinosus; chicken and turkey, thigh.

[c]White muscle as follows: mutton and beef, none; pork, white portion of semitendinosus; chicken and turkey, breast.

TABLE 6-2

Fatty-Acid Composition of Triglyceride and Phospholipid Fractions from Lean Beef and Pork[a]

Fatty Acid	Triglycerides, %		Phospholipids, %	
	Beef	Pork	Beef	Pork
Saturated				
Capric	0.1	0.1	-	-
Lauric	0.1	0.2	-	-
Myristic	2.2	1.2	2.6	2.0
Stearic	16.9	11.6	15.6	11.0
Palmitic	27.5	23.9	13.2	20.0
Unsaturated				
Palmitoleic	4.7	7.4	2.2	2.3
Oleic	41.3	45.2	21.2	16.2
Tetradecadienoic	0.6	-	1.3	0.6
Linoleic	4.4	8.7	20.2	27.9
Linolenic	1.1	1.6	1.8	1.0
Eicosatrienoic	-	-	1.8	1.6
Arachidonic	0.1	0.1	19.2	16.3
Docosadienoic	-	-	-	0.9
Total saturated acids	46.8	37.0	31.4	33.0
Total monounsaturated acids	47.1	52.6	24.3	18.8
Total fatty acids with > 1 unsaturated bond	6.2	10.4	44.3	48.2

[a]Adapted from Hornstein *et al*., 1961. Reprinted with permission from Journal of Food Science **26**:581, 1961. Copyright © by the Institute of Food Technologists.

Numerous compounds, including heme and nonheme iron, may act as catalysts of lipid oxidation (Liu and Watts, 1970; Wills, 1966). Muscle tissue contains a considerable amount of iron bound to proteins. Myoglobin is an oxygen-storage protein in muscle cells resembling hemoglobin. Muscle tissue probably also contains residues of hemoglobin from blood. In addition, cells contain cytochromes. All these proteins contain the prosthetic group, heme, which has an iron atom at its center. The iron atom by itself promotes autoxidation of fats; in the case of heme, the whole iron-heme molecule may participate in the oxidation reaction.

Although schemes for catalysis of lipid oxidation have been proposed (Tappel, 1962; Tarladgis, 1961), the situation in meat is difficult to ascertain (Kendrick and Watts, 1969). There is recent evidence that nonheme iron is the major prooxidant in cooked meat (Igene *et al.*, 1979; Love and Pearson, 1974a,b; Sato and Hegarty, 1971). Igene *et al.* (1979) further demonstrated that nonheme iron is released from heme pigments as a consequence of cooking, or by treatment with hydrogen peroxide, thus accelerating lipid oxidation. These results verified the report by Haurowitz *et al.* (1941) that the prooxidation effect of hemin or hemoglobin on linoleic and linolenic acids is due to release of inorganic iron. It is possible that other polyvalent cations are prooxidants in meat and play a role in WOF. However, Sato and Hegarty (1971) demonstrated that cupric salts actually inhibited WOF, apparently through the reaction of free radicals with cupric ions. Thus, nonheme iron appears to be the major prooxidant in development of WOF (Igene *et al.*, 1979).

MEASUREMENT OF LIPID OXIDATION IN MEAT SYSTEMS

There are many experimental techniques for determining the extent of lipid oxidation in biologic systems, from simple sensory evaluations to chemical and physical measurements. Chemical methods include determinations of peroxide, thiobarbituric acid, total and volatile carbonyls, and oxirane. Physical methods include the use of conjugated diene, fluorescence, infrared spectroscopy, polarography, gas chromatography, and refractometry. These methods have been reviewed by Sherwin (1968), Erickson and Bowers (1976), Gray (1978), and Logani and Davies (1980). Two of the more commonly used methods are reviewed below.

Thiobarbituric Acid Test

One of the most commonly used methods for assessing lipid oxidation in meat and other biologic systems is the 2-thiobarbituric acid (TBA) method. The extent of oxidative rancidity is usually expressed as a TBA number (milligrams of malonaldehyde per kilogram of sample) by comparing the optical density of the TBA-malonaldehyde colored complex with that of standards prepared from 1,1,3,3-tetraethoxypropane. In acidic solution, the acetal is quantitatively hydrolyzed to malonaldehyde. The method is based on the development and measurement of a red pigment formed by the condensation of one molecule of malonaldehyde and two molecules of TBA (Sinnhuber and Yu, 1958). The acid is added to inhibit the possible reaction of malonaldehyde with food constituents.

Because of its reactivity, most of the malonaldehyde in foods is bound to other food components, and very little is in the free form. Kwon and Watts (1963) thus proposed the term "distillable malonaldehyde" for use in describing the TBA test. They concluded (1964) that malonaldehyde has the capacity to enolize from its diketo form (I) to its enolate anion (II), which is not volatile. A third volatile chelated form (III) is also possible.

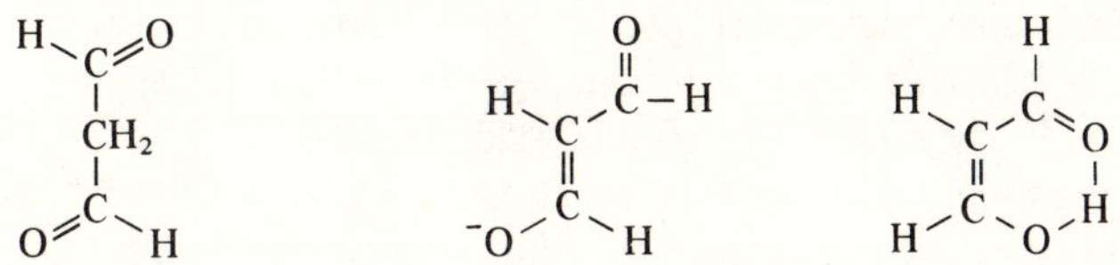

(I) (II) (III)

Kwon and Watts (1964) indicated that, in aqueous solution, almost all (96%) of the malonaldehyde is in the enolic form (II) and that the proportions of the various forms present depend on pH. The enolic form occurs at a pH of 7, and chelated form III predominates at a pH less than 3. Therefore, Kwon and Watts (1964) stated that maximal volatization of free, preformed malonaldehyde occurs at a pH less than 3.

The TBA test has been performed in various ways, of which the two most common are as follows: (1) The TBA reagent in strong acid is added directly to the food product, and the whole mixture is heated in a water bath until maximal color is developed; the color complex is extracted with a solvent and measured spectrophotometrically (Sinnhuber *et al.*, 1958; Yu and Sinnhuber, 1957). (2) The food product is *first* steam-distilled from an acidic solution, and a portion of the distillate is mixed with the TBA reagent; this mixture is heated and the resulting color complex is measured directly in a spectrophotometer (Tarladgis *et al.*, 1960). Both methods require the presence of strong acid (pH, 0.9-1.5), to liberate malonaldehyde from some precursors and to catalyze the condensation of the aldehyde with the TBA reagent.

The TBA test has been subject to some criticism (Gray, 1978; Logani and Davies, 1980; Sinnhuber and Yu, 1977). Yu and Sinnhuber (1964) pointed out the importance of using noncontaminated reagents for the test, to prevent side reactions that would give misleading results. Some compounds react with TBA to yield colored complexes with absorption maximums between 450 and 530 nm (Marcuse and Johansson, 1973), and the color intensity can be such that it would interfere with absorption at 530 nm and cause a higher than normal value. Buttkus and Bose (1972) found that malonaldehyde-aromatic amine reaction products yielded lower TBA values than expected, because they were relatively stable in the presence of acid and heat. Dillard and Tappel (1971) reported that, in peroxidizing microsomes, the TBA value stabilized, whereas oxygen absorption continued. Wills (1966) demonstrated that TBA values increased much more than oxygen uptake in the presence of ascorbate and postulated that partially oxidized fatty acids in a tissue homogenate could break down without further oxygen absorption and thus yield additional TBA-reacting products. It is also unclear whether malonaldehyde is the ultimate product in lipid oxidation that reacts with TBA or some other reactive material generates malonaldehyde under the conditions of the TBA test (Logani and Davies, 1980). Pryor *et al.* (1976) have suggested that the prostaglandin endoperoxides may be precursors of malonaldehyde under the test conditions.

The TBA test has been modified for measuring the extent of lipid oxidation in meat samples containing nitrite. Hougham and Watts (1958) reported that the presence of nitrite at 200 mg/kg decreased TBA values by 20-30%, but nitrite at less than 100 mg/kg did not interfere with the test. Zipser and Watts (1962) stated that small amounts of nitrite ion can substantially reduce TBA numbers in rancid meat, with the reduction increasing linearly with nitrite concentration. Nitrite interference with the TBA test takes place during the distillation step and is believed to be due to nitrosation of malonaldehyde. Therefore, sulfanilamide is usually added, to bind the nitrite through diazonium salt formation and thus yield accurate TBA readings for samples containing nitrite.

Total-Carbonyl Determinations

An alternative approach for measuring lipid oxidation in meat systems is to measure the carbonyl compounds formed by the degradation of the hydroperoxides (Gray, 1978). The most reliable and widely used of the analytic methods is the one described by Henick *et al.* (1954). The procedure is based on the formation of 2,4-dinitrophenylhydrazones of carbonyl compounds in the presence of trichloroacetic acid catalyst. However, the method has been criticized on the grounds that hydroperoxides may decompose under the experimental conditions (Lea, 1962).

MacDonald *et al.* (1980) used the high-performance liquid-chromatographic procedure of Selim (1977) to measure the hexanal and 2,4-decadienal concentrations in hams treated with 2.5% salt and various concentrations of nitrite, butlylated hydroxytoluene (BHT) and citric acid.

Significantly lower ($p < 0.05$) concentrations of these aldehydes were found in meat treated with nitrite, BHT, or citric acid than in samples treated only with salt. Previous studies by Cross and Ziegler (1965) and Swain (1972) revealed lower concentrations of hexanal in the volatiles from nitrite-treated hams than in those from meat samples free of nitrite.

AGENTS INHIBITING LIPID OXIDATION

Nitrite

Nitrite has been shown to inhibit lipid oxidation (WOF) in cooked meat and meat products. Sato and Hegarty (1971) reported that lipid oxidation, as measured by TBA values, was eliminated in cooked ground beef by adding sodium nitrite at 2,000 mg/kg. Sodium nitrite at 50 mg/kg reduced TBA values by 65% in their studies as well. They also suggested that the substance(s) responsible for initiating WOF in cooked meats was water-soluble, inasmuch as beef muscle that had been thoroughly extracted with water did not develop WOF.

Younathan and Watts (1959) studied lipid oxidation in cured and uncured refrigerated cooked pork and found the highest TBA values in uncured samples at all storage periods over 2 wk. Similar studies on comminuted pork were carried out by Hadden *et al.* (1975), who reported that storage time at $3^{o} \pm 2^{o}C$ and the absence of nitrite significantly increased TBA values ($p < 0.01$). This trend was also observed when the products were stored at $-29^{o} \pm 2^{o}$. Samples of cooked beef, pork, and chicken prepared with and without sodium nitrite (156 mg/kg) have been evaluated by the TBA test and by sensory-panel scores at the beginning of storage and after storage for 48 h at $4^{o}C$ (Fooladi *et al.*, 1979). Added nitrite reduced TBA values by a factor of 2 in beef and chicken samples and a factor of 5 in pork. Sensory-panel scores also indicated a protective effect of added nitrite.

The effects of various concentrations of sodium nitrite (50, 200, and 500 mg/kg) on the oxidation stability of cooked hams have been studied recently by MacDonald *et al.* (1980). Their data (Figure 6-1) indicated a significant reduction in TBA values for pork cured with sodium nitrite.

As indicated previously, Sato and Hegarty (1971) and Love and Pearson (1974a,b) have presented data that suggest that nonheme iron is the major prooxidant in cooked meat. Igene *et al.* (1979) further investigated the influence of heme pigments, nonheme iron, and nitrite on development of WOF in cooked meat and showed that removal of meat pigments and addition of sodium nitrite at 156 mg/kg significantly ($p < 0.001$) inhibited lipid oxidation in samples of cooked meat (beef and dark and white chicken meat). Taste-panel evaluation confirmed the inhibitory effects of removal of heme pigments and addition of nitrite on the development of WOF.

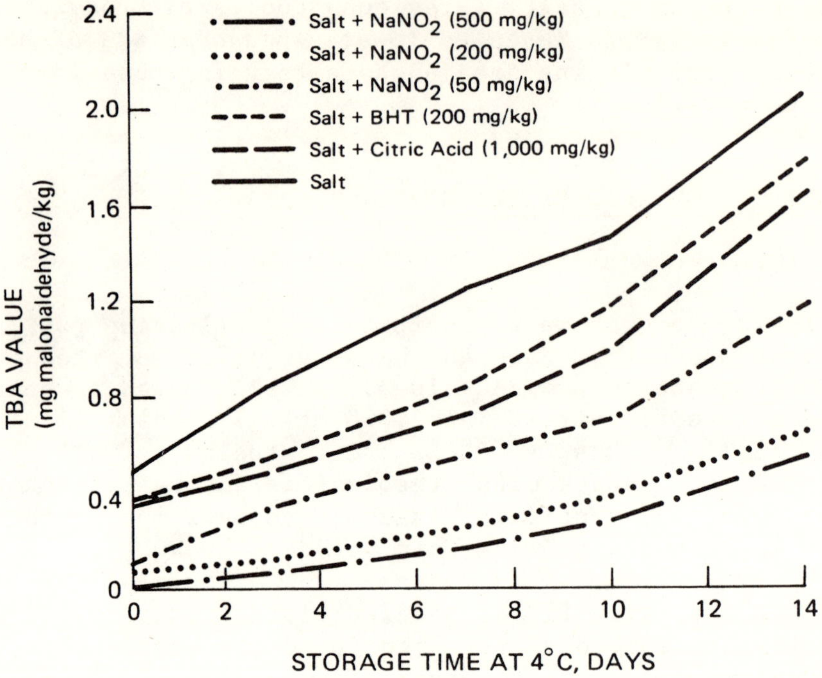

FIGURE 6-1. Effect of curing agents, antioxidants, and metal chelating compounds on TBA values of pork stored at 4°C. Adapted from MacDonald et al., 1980. Reprinted with permission from Journal of Food Science 45:899-892, 1980. Copyright © by Institute of Food Technologists.

The percentage of bound heme iron in fresh meat pigment extract was slightly greater than 90%, whereas that of free nonheme iron was less than 10%. Cooking, however, released a significant amount of nonheme iron from the bound heme pigments and accelerated lipid oxidation in cooked meat. Thus, it was concluded that the increased rate of lipid oxidation in cooked meat was due to release of nonheme iron during cooking. This study verified the report by Haurowitz *et al.* (1941) that the prooxidant effect of hemin or hemoglobin on linoleic and linolenic acids was due to release of inorganic iron.

The mechanism by which nitrite minimizes WOF in cooked meats is not thoroughly understood, although it has been suggested that nitrite may either stabilize the lipid components of the membranes or inhibit the natural prooxidants in muscle (Pearson *et al.*, 1977). Clearly, further studies are needed to elucidate fully the antioxidant mechanism of nitrite in cooked meats.

Some nitrite derivatives in meat also have antioxidant properties. *S*-Nitrosocysteine, a compound generated during the curing of meat, has been shown to act as an antioxidant in both an aqueous linoleate model system and ground cooked turkey meat (Kanner and Juven, 1980). Kanner *et al.* (1980) reported the antioxidant activity of nitric oxide myoglobin (Mb-NO) in linoleate and β-carotene-linoleate aqueous model systems. The specific antioxidant activity of Mb-NO was maintained even in the presence of prooxidants, such as heme proteins and lipoxygenase.

Other Antioxidants Tested in Meat or Fish

Sato and Hegarty (1971) tested a variety of compounds for their ability to inhibit lipid oxidation, as measured by TBA values (Table 6-3). The most active compounds were sodium ethylene diaminetetraacetic acid (EDTA), sodium tripolyphosphate, sodium hexametaphosphate, sodium citrate, sodium ascorbate, butylated hydroxyanisole (BHA), and butylated hydroxytoluene (BHT). However, only the last two compounds were effective at low concentrations (100 mg/kg).

Recently, MacDonald *et al.* (1980) tested citric acid and BHT for antioxidant effects and compared the results with those of nitrite at various concentrations (Figure 6-1). They reported that citric acid (1,000 mg/kg) and BHT (200 mg/kg) were less active than sodium nitrite at the lowest concentration tested (50 mg/kg).

Sweet (1973) reported that untreated (control) samples of salmon or trout developed TBA values indicative of marked rancidity in less than 24 h at 5°C. Antioxidants increased the length of storage before development of rancidity, as measured by TBA values. Combinations of either BHA or *tert*-butylhydroquinone (TBHQ) with EDTA or citric acid were the most effective inhibitors of lipid oxidation in the fish samples tested.

TABLE 6-3

Effect of Various Reagents on TBA Value of Cooked Ground Beef Stored 2 Days at 4°C[a]

Reagent	Concentration of Reagent, mg/kg	Inhibition of Lipid Oxidation, %
2Na-EDTA	2,500	92
Na tripolyphosphate	5,000	95
Na hexametaphosphate	5,000	88
Na pyrophosphate	5,000	82
Na citrate	5,000	12
Na ascorbate	5,000	82
Na ascorbate	50	<1
BHT	100	81
BHA	100	85
$NaNO_2$	2,000	99
$NaNO_2$	50	65
$ZnSO_4$	50	0 (2% promotion)
$CaSO_4$	50	1
$MgSO_4$	50	12
$MnSO_4$	50	0 (30% promotion)
$CuSO_4$	50	18
$CuSO_4$	100	34
$CuCl_2$	150	94
$AgNO_3$	150	23
$HgCl_2$	150	32
Iron powder	300	0 (75% promotion)
$FeCl_3$	100	2
$FeCl_2$	100	14
KCN	100	36
L-Cysteine	5,000	57
$NaHSO_3$	2,000	58

[a]Percent inhibition calculated by using data from Sato and Hegarty (1971). Reprinted with permission from Journal of Food Science 36:1098, 1971.

The addition of some phosphates--including pyrophosphate, tripolyphosphate, and hexametaphosphate--has been shown to protect cooked meats from lipid oxidation (Tims and Watts, 1958). These effects were confirmed by Sato and Hegarty (1971), who demonstrated that all three compounds markedly reduced the TBA values in cooked ground beef stored at 2°C (Table 6-3). The mechanism by which phosphates prevent lipid oxidation appears to be related to their ability to sequester metal ions, particularly ferrous ions, which are the major prooxidants in meat systems (Love and Pearson, 1974a,b).

Ascorbic acid at concentrations below 1,000 mg/kg can enhance the development of WOF, as shown by increased TBA values (Sato and Hegarty, 1971; Tims and Watts, 1958). At higher concentrations (5,000-10,000 mg/kg), however, ascorbic acid inhibits lipid oxidation (Sato and Hegarty, 1971; Table 6-3), probably by upsetting the balance between ferrous and ferric ions or by acting as an oxygen scavenger.

It has also been demonstrated by Tims and Watts (1958) that ascorbic acid acts synergistically with phosphates to protect against rancidity. Similar results were obtained by Sato and Hegarty (1971). These studies indicated that ascorbic acid and phosphates in combination may have an important synergistic action in preventing the development of oxidation in cured meats, as suggested by Chang and Watts (1949).

The presence or absence of tocopherols (e.g., α-tocopherol or vitamin E) in animal tissues can influence rancidity and presumably the development of WOF (Pearson et al., 1977). Studies thus far have been based on the effects in meat of animals fed tocopherol; however, tocopherol added directly to meats has not been evaluated as an antioxidant.

The antioxidant effect on rat body tissues after dietary supplementation with tocopherol was shown by Barnes et al. (1943). The dietary application of tocopherol to stabilize pork fat has been investigated by Watts et al. (1946) and Hvidsten and Astrup (1963). Only a slight improvement in keeping quality was achieved by this method, and the improvement was thought to be too small for practical importance. Tasai et al. (1978) further studied the extent of improvement in the oxidative stability of pork from pigs fed rations containing α-tocopherol acetate and ascorbic acid. The oxidative stability of the pork tissues was unaffected by ascorbic acid supplementation of the diet; however, the stability of adipose tissues was significantly increased ($p < 0.01$) by α-tocopherol acetate supplementation. Supplementation with α-tocopherol acetate at 100 mg/kg in the feed appeared to be optimal for improvement of the oxidative stability of pork.

Dietary supplementation of feed with vitamin E has been shown to be effective in other animal tissues. Webb et al. (1972, 1974) reported that vitamin E supplementation improved the TBA values of precooked frozen turkey and chicken parts, but they were unable to establish a relationship between sensory-panel scores and TBA values. Ellis *et al.* (1974) found that the keeping quality of veal was improved by feeding calves vitamin E with milk from cows that had been fed protected safflower oil and, after weaning, by feeding calves protected safflower oil directly.

Sato *et al.* (1973) reported that the overcooking of meat protects against the development of WOF by producing compounds with antioxidant activity. They studied the reaction that is responsible for producing antioxidant activity during the heating of meat and concluded that the substances demonstrating antioxidant properties were the result of the Maillard reaction, i.e., the heat-catalyzed interaction between amino acids or proteins and carbohydrates. They also found that reductic acid, maltol, and products of the amino-sugar reaction were effective inhibitors of development of WOF in cooked ground beef. The antioxidant nature of products of the Maillard reaction has been discussed more fully by Eichner (1980).

Antioxidants Not Tested in Meat Systems

Flavonoids (Polyphenols). Flavonoids have strong antioxidant properties and occur widely in the plant kingdom--in fruits, vegetables, leaves, and flowers. They exist primarily in water-soluble forms as glycosides or esterified with quinic acid. They can also occur as part of complex lipid structures.

Oil-bearing seeds contain substantial proportions of these compounds. Extracts of soybeans, for example, have strong antioxidant properties due in large part to a mixture of flavonoids. The most active components of the mixture are the cinnamic acid derivatives of chlorogenic acids (Pratt, 1980). Although these compounds exist in the natural state in water-soluble forms, they could easily be rendered lipid-soluble by the substitution of long-chain alcohols or fatty acids.

A number of chemically pure flavones have been tested for their ability to retard the oxidation of lard. Four of them increase the shelf-life of lard by a factor of 4-10. They are, in order of increasing effectiveness, quercetin, robinetin, myricetin, and gossypetin (Mehta and Seshadvi, 1959).

Wood smoke has been used as a meat preservative for centures. It is generally believed that the observed antioxidant activity of wood smoke is due to the presence of polyphenols in the smoke.

Spices. Herbs and spices have traditionally been used as antioxidants. In India, for example, capsicum, onion, and garlic are used in the butter-like product, ghee, to prolong stability and increase palatability. The antioxidant activity of specific spices has been reviewed by Dugan (1980).

CONCLUSIONS AND RECOMMENDATIONS

Cured meats that have been cooked or comminuted are highly susceptible to the development of rancidity caused by lipid oxidation. Nitrite, at concentrations currently used in cured-meat products, has been shown to inhibit lipid oxidation in several cooked-meat products. The antioxidant effect of nitrite is particularly important in comminuted products, in which air may be incorporated during manufacture, and there is some evidence that it is more effective in pork products than in chicken and beef products. However, the mechanism of action of nitrite and the minimal concentration required in cured meats to inhibit lipid oxidation sufficiently for the probable period before such products are consumed are now known.

Agents other than nitrite have demonstrated antioxidant activity in meat products. For example, BHT and citric acid have been reported by several research groups to be effective in inhibiting lipid oxidation in ground beef. However, neither agent appears to be as effective as sodium nitrite at 50 mg/kg. Tocopherols may inhibit lipid oxidation in meat from animals fed these compounds. Combinations of TBHQ or BHA with EDTA or citric acid have been shown to inhibit lipid oxidation in fish. Several agents have also been found to inhibit lipid oxidation in other food systems.

The Committee makes the following recommendations:

Further work is needed to define the mechanism(s) by which nitrite inhibits lipid oxidation and to determine the minimal concentrations of nitrite that will produce the necessary inhibition of lipid oxidation. Investigations are also needed to determine whether the reductions in lipid oxidation achieved at lower concentrations of nitrite are sufficient to prevent warmed-over flavor and rancidity and to guard against the possible health hazards associated with some products of lipid oxidation for the probable period of product storage before consumption.

Further testing in meat systems--and in a variety of cured-meat products--of other agents found to inhibit lipid oxidation in meats or other food systems should be conducted before any conclusion can be drawn concerning their applicability as a substitute for nitrite.

REFERENCES

Ames, B. N., M. C. Hollstein, and R. Cathcart. In press. Lipid peroxidation and oxidative damage to DNA. In K. Yagi, ed. Lipid Peroxide in Biology and Medicine. Academic Press, New York.

Anonymous. 1981. Bacon. Cons. Reports 46:496-501.

Apaja, M. 1980. Evaluation of toxicity and carcinogenicity of malonaldehyde. Acta Univ. Oul. D55. Anat. Pathol. Microbiol. 8:1-61.

Barnes, R. H., W. O. Lundberg, H. T. Hanson, and G. O. Burr. 1943. The effect of certain dietary ingredients on the keeping quality of body fat. J. Biol. Chem. 149:313-322.

Benedict, R. C. 1980. Biochemical basis for nitrite-inhibition of *Clostridium botulinum* in cured meat. J. Food Prot. 43:877-891.

Brooks, B. R., and O. L. Klamerth. 1968. Interaction of DNA with bifunctional aldehydes. Eur. J. Biochem. 5:178-182.

Buttkus, H., and R. J. Bose. 1972. Amine-malonaldehyde condensation products and their relative color contribution in the thiobarbituric acid test. J. Am. Oil Chem. Soc. 49:440-443.

Chang, I., and B. M. Watts. 1949. Antioxidants in the hemoglobin-catalyzed oxidation of unsaturated fats. Food Technol. 3:332-336.

Cross, C. K., and P. Ziegler. 1964. A comparison of the volatile fractions from cured and uncured meat. J. Food Sci. 30:610-614.

Deuel, H. J., Jr. 1955. The Lipids: Their Chemistry and Biochemistry, Volume II. Interscience Publishers, Inc., New York.

Dillard, C. J., and A. L. Tappel. 1971. Fluorescent products of lipid peroxidation of mitochondria and microsomes. Lipids 6:715-721.

Dugan, L. R. 1980. Natural antioxidants. Pp. 261-282 in M. G. Simic and M. Karel, eds. Autoxidation in Food and Biological Systems. Plenum Press, New York and London.

Eichner, K. 1980. Antioxidative effect of maillard reaction intermediates. Pp. 367-385 in M. G. Simic and M. Karel, eds. Autoxidation in Food and Biological Systems. Plenum Press, New York and London.

Ellis, R., W. I. Kimoto, J. Bitman, and L. F. Edmondson. 1974. Effect of induced high linoleic acid and tocopherol content on the oxidative stability of rendered veal fat. J. Am. Oil Chem. Soc. 51:4-7.

Erickson, D. R., and R. H. Bowers. 1976. Objective determination of fat stability in prepared foods. Pp. 133-144 in Proceedings of a Symposium: Objective Methods for Food Evaluation. National Academy of Sciences, Washington, D.C.

Fooladi, M. H., A. M. Pearson, T. H. Coleman, and R. A. Merkel. 1979. The role of nitrite in preventing development of warmed-over flavour. Food Chem. 4:282-292.

Gaddis, A. M., R. Ellis, and G. T. Currie. 1961. Carbonyls in oxidizing fat. V. The composition of neutral volatile monocarbonyl compounds from autoxidized oleate, linoleate, linolenate esters, and fats. J. Am. Oil Chem. Soc. 38:371-375.

Giam, I., and L. R. Dugan, Jr. 1965. The fatty acid composition of free and bound lipids in freeze-dried meats. J. Food Sci. 30:262-265.

Gray, J. I. 1978. Measurement of lipid oxidation: A review. J. Am. Oil Chem. Soc. 55:539-546.

Hadden, J. P., H. W. Ockerman, V. R. Cahill, N. A. Parrett, and R. J. Borton. 1975. Influence of sodium nitrite on the chemical and organoleptic properties of comminuted pork. J. Food Sci. 40:626-630.

Haurowitz, F., P. Schwerin, and M. M. Yenson. 1941. Destruction of hemin and hemoglobin by the action of unsaturated fatty acids and oxygen. J. Biol. Chem. 140:353-359.

Henick, A. S., M. F. Benca, and J. H. Mitchell, Jr. 1954. Estimating carbonyl compounds in rancid fats and foods. J. Am. Oil Chem. Soc. 31:88-91.

Hilditch, T. P., and P. N. Williams. 1964. The Chemical Constitution of Natural Fats, 4th Ed. John Wiley & Sons Inc., New York.

Hornstein, I. 1967. Flavor of red meats. Pp. 228-250 in H. W. Schultz, E. A. Day, and L. M. Libbey, eds. Symposium on Foods: The Chemistry and Physiology of Flavors. AVI Publishing Company, Inc., Westport, Connecticut.

Hornstein, I., P. F. Crowe, and W. L. Sulzbacher. 1960. Constituents of meat flavor: Beef. J. Agric. Food Chem. 8:65-67.

Hornstein, I., P. F. Crowe, and M. J. Heinberg. 1961. Fatty acid composition of meat tissue lipids. J. Food Sci. 26:581.

Horvat, R. J., W. H. McFadden, H. Ng, W. G. Lane, A. Lee, R. E. Lundin, J. R. Scherer, and A. D. Shepherd. 1969. Identification of some acids from autoxidation of methyl linoleate. J. Am. Oil Chem. Soc. 46:94-96.

Hougham, D., and B. M. Watts. 1958. Effect of variations in curing salts on oxidative changes in radiation sterilized pork. Food Technol. 12:681-684.

Hvidsten, H., and H. Astrup. 1963. The effect of vitamin E on the keeping quality and flavour of pork. Acta Agric. Scand. 13:259-270.

Igene, J. O., J. A. King, A. M. Pearson, and J. I. Gray. 1979. Influence of heme pigments, nitrite and none-heme iron on development of warmed-over flavor (WOF) in cooked meat. J. Agric. Food Chem. 27:838-842.

Kanner, J., and B. J. Juven. 1980. S-Nitrosocysteine as an antioxidant, color-developing, and anticlostridial agent in comminuted turkey meat. J. Food Sci. 45:1105-1112.

Kendrick, J., and B. M. Watts. 1969. Acceleration and inhibition of lipid oxidation by heme compounds. Lipids 4:454-458.

Kikugawa, K., K. Tsukuda, and T. Kurechi. 1980. Studies on peroxidized lipids. I. Interaction of malondialdehyde with secondary amines and its relevance to nitrosamine formation. Chem. Pharm. Bull. 28:3323-3331.

Koch, D. E., A. M. Pearson, W. T. Magee, J. A. Hoefer, and B. S. Schweigert. 1968. Effect of diet on the fatty acid composition of pork fat. J. Anim. Sci. 27:360-365.

Kurechi, T., K. Kikugawa, and M. Ozawa. 1980. Effect of malondialdehyde on nitrosamine formation. Food Cosmet. Toxicol. 18: 119-122.

Kwon, T. W., and B. M. Watts. 1963. Determination of malonaldehyde by ultraviolet spectrophotometry. J. Food Sci. 28:627-630.

Kwon, T.-W., and B. M. Watts. 1964. Malonaldehyde in aqueous solution and its role as a measure of lipid oxidation in foods. J. Food Sci. 29:294-302.

Lea, C. H. 1962. The oxidative deterioration of food lipids. Pp. 3-28 in H. W. Schultz, E. A. Day, and R. C. Sinnhuber, eds. Symposium on Foods: Lipids and Their Oxidation. AVI Publishing Company, Westport, Connecticut.

Link, B. A., R. W. Bray, R. G. Cassens, and R. G. Kauffman. 1970a. Fatty acid composition of bovine skeletal muscle lipids during growth. J. Anim. Sci. 30:726-731.

Link, B. A., R. W. Bray, R. G. Cassens, and R. G. Kauffman. 1970b. Fatty acid composition of bovine subcutaneous adipose tissue lipids during growth. J. Anim. Sci. 30:722-725.

Liu, H.-P., and B. M. Watts. 1970. Catalysts of lipid peroxidation in meats. 3. Catalysts of oxidative rancidity in meats. J. Food Sci. 35:596-598.

Logani, M. K., and R. E. Davies. 1980. Lipid oxidation: Biologic effects and antioxidants--A review. Lipids 15:485-495.

Love, J. D., and A. M. Pearson. 1971. Lipid oxidation in meat and meat products--A review. J. Am. Oil Chem. Soc. 48:547-549.

Love, J. D., and A. M. Pearson. 1974a. Metmyoglobin and nonheme iron as prooxidants in egg-yolk phospholipid dispersions and cooked meat. J. Agric. Food Chem. 24:494-498.

MacDonald, B., J. I. Gray, Y. Kakuda, and M. L. Lee. 1980. Role of nitrite in cured meat flavor: Chemical analysis. J. Food Sci. 45:889-892.

Marcuse, R., and L. Johansson. 1973. Studies on the TBA test for rancidity grading: II. TBA reactivity of different aldehyde classes. J. Am. Oil Chem. Soc. 50:387-391.

Marnett, L. J., and M. A. Tuttle. 1980. Comparison of the mutagenicities of malondialdehyde and the side products formed during its chemical synthesis. Cancer Res. 40:276-282.

Mehta, A. C., and T. R. Seshadvi. 1959. Flavonoids as antioxidants. J. Sci. Ind. Res. (India) 18B:24.

Mukai, F. H., and B. D. Goldstein. 1976. Mutagenicity of malonaldehyde, a decomposition product of peroxidized polyunsaturated fatty acids. Science 191:868-869.

National Academy of Sciences. 1981. The Health Effects of Nitrate, Nitrite, and N-Nitroso Compounds. National Academy Press, Washington, D.C. 544 pp.

O'Keefe, P. W., G. H. Wellington, L. R. Mattick, and J. R. Stouffer. 1968. Composition of bovine muscle lipids at various carcass locations. J. Food Sci. 38:188-192.

Orme, L. E., A. M. Pearson, L. J. Bratzler, and W. T. Magee. 1958. Specific gravity as an objective measure of marbling. J. Anim. Sci. 17:693-698.

Pearson, A. M., J. D. Love, and F. B. Shorland. 1977. "Warmed-over" flavor in meat, poultry, and fish. Adv. Food Res. 23:1-74.

Pratt, D. E. 1980. Natural antioxidants of soybeans and other oil-seeds. pp. 283-293 in M. G. Simic and M. Karel, eds. Antioxidation in Foods and Biological Systems. Plenum Press, New York and London.

Pryor, W. A., J. P. Stanley, and E. Blair. 1976. Autoxidation of polyunsaturated fatty acids: II. A suggested mechanism for the formation of TBA-reactive materials from prostaglandin-like endoperoxides. Lipids 11:370-379.

Sato, K., and G. R. Hegarty. 1971. Warmed-over flavor in cooked meats. J. Food Sci. 36:1098-1102.

Sato, K., G. R. Hegarty, and H. K. Herring. 1973. The inhibition of warmed-over flavor in cooked meats. J. Food Sci. 38:398-403.

Selim, S. 1977. Separation and quantitative determination of traces of carbonyl compounds as their 2,4-dinitrophenylhydrazones by high-pressure liquid chromatography. J. Chromatogr. 136:271-277.

Shamberger, R. J., T. L. Andreone, and C. E. Wills. 1974. Antioxidants and cancer. IV. Initiating activity of malonaldehyde as a carcinogen. J. Natl. Cancer Inst. 53:1404-1409.

Sherwin, E. R. 1968. Methods for stability and antioxidant measurement. J. Am. Oil Chem. Soc. 45:632A.

Sinnhuber, R. O., and T. C. Yu. 1977. The 2-thiobarbituric acid reaction, an objective measure of the oxidative deterioration occurring in fats and oils. J. Jpn. Oil Chem. Soc. [Yukagaku] 26(5):259-267.

Sinnhauber, R. O., T. C. Yu, and T. C. Yu. 1958. Characterization of the red pigment formed in the 2-thiobarbituric acid determination of oxidative rancidity. Food Res. 23:626-634.

Siu, G. M., and H. H. Draper. 1978. A survey of the malonaldehyde content of retail meats and fish. J. Food Sci. 43:1147-1149.

Summerfield, F. W., and A. L. Tappel. 1978. Enzymatic synthesis of malonaldehyde. Biochem. Biophys. Res. Commun. 82:547-552.

Swain, J. W. 1972. Volatile Flavor Constituents of Pork Cured With and Without Nitrites. Ph.D. Dissertation, University of Missouri, Columbia, Missouri.

Sweet, C. W. 1973. Activity of antioxidants in fresh fish. J. Food Sci. 38:1260-1261.

Tappel, A. L. 1962. Hematin compounds and lipoxidase as biocatalysts. Pp. 122-138 in H. W. Schultz, E. A. Day, and R. O. Sinnhuber, eds. Symposium on Foods: Lipids and Their Oxidation. AVI Publishing Company, Westport, Connecticut.

Tarladgis, B. G. 1961. An hypothesis for the mechanism of the heme catalyzed lipid oxidation in animal tissues. J. Am. Oil Chem. Soc. 38:479-483.

Tarladgis, B. G., B. M. Watts, M. T. Younathan, and L. Dugan, Jr. 1960. A distillation method for the quantitative determination of malonaldehyde in rancid foods. J. Am. Oil Chem. Soc. 37:44-48.

Tims, M. J., and B. M. Watts. 1958. Protection of cooked meats with phosphates. Food Technol. 12:240-243.

Tsai, T. C., G. H. Wellington, and W. G. Pond. 1978. Improvement in the oxidative stability of pork by dietary supplementation of swine rations. J. Food Sci. 43:193-196.

Watts, B. M. 1962. Meat products. Pp. 202-214 in H. W. Schultz, E. A. Day, and R. O. Sinnhuber, eds. Symposium on Foods: Lipids and Their Oxidation. AVI Publishing Company, Westport, Connecticut.

Watts, B. M., T. J. Cunha, and R. Major. 1946. Effect of feeding and injecting hogs with tocopherols on the susceptibility of pork fat to rancidity. Oil Soap (Chicago) 23:254.

Webb, J. E., C. C. Brunson, and J. D. Yates. 1972. Effects of feeding antioxidants on rancidity development in pre-cooked, frozen broiler parts. Poult. Sci. 51:1601-1605.

Webb, J. E., C. C. Brunson, and J. D. Yates. 1974. Effects of dietary fat and dl-α-tocopheryl on stability characteristics of precooked frozen broiler parts. J. Food Sci. 39:133-136.

Wills, E. D. 1966. Mechanisms of lipid peroxide formation in animal tissues. Biochem. J. 99:667-676.

Wilson, B. R., A. M. Pearson, and F. B. Shorland. 1976. Effect of total lipids and phospholipids on warmed-over flavor in red and white muscle from several species as measured by thiobarbituric acid. J. Agric. Food Chem. 24:7-10.

Younathan, M. T., and B. M. Watts. 1959. Relationship of meat pigments to lipid oxidation. Food Res. 24:728-734.

Younathan, M. T., and B. M. Watts. 1960. Oxidation of tissue lipids in cooked pork. Food Res. 25:538-543.

Yu, T. C., and R. O. Sinnhuber. 1957. 2-Thiobarbituric acid method for the measurement of rancidity in fishery products. Food Technol. 11:104-108.

Yu, T. C., and R. O. Sinnhuber. 1964. Further observations on the 2-thiobarbituric acid method for the measurement of oxidative rancidity. J. Am. Oil Chem. Soc. 41:540-543.

Zipper, M. W., and B. M. Watts. 1962. A modified 2-thiobarbituric acid (TBA) method for the determination of malonaldehyde in cured meats. Food Technol. 16(7):102-104.

CHAPTER 7

PRODUCTION OF CURED-MEAT COLOR IN RED-MEAT PRODUCTS

Page

CHAPTER 7

PRODUCTION OF CURED-MEAT COLOR IN RED-MEAT PRODUCTS

The bright appearance created by the oxymyoglobin in muscle tissue of fresh red meat and the reddish-pink hue of denatured nitrosylmyohemochrome in cured meat products are attributes recognized by the consumer. Although the color of a meat product does not necessarily predict good texture and flavor, the shopper appears to make such an association (Giddings, 1977a; Jeremiah *et al*., 1972). This apparent consumer preference for specific product colors is critical to deliberations on the need for nitrite or an alternative. Various questions need to be answered for informed decisions to be made.

- To what extent do consumers actually prefer particular colors in cured products?
- To what extent is such a desire learned, and how rapidly might it be influenced by educational programs or decreased availability?
- To what extent is color a cue for other sensory judgments?

For an adequate assessment of alternatives to the current use of nitrite for color fixation in cured red meat products, information on a number of questions is desirable. These include:

- What is the minimal concentration of nitrite or other agent that will produce color, and is there a dose-response relationship for the effect?
- What is the mechanism of color formation, and to what extent do other components of the conventional cure (e.g., salt, ascorbate, sucrose) or processing treatments (e.g., smoking) contribute to or affect the color produced by nitrite or alternatives?
- What factors (e.g., light, temperature, and gaseous atmosphere) affect the stability of the color formed by current nitrite cures, by reduced amounts of nitrite, or by alternatives?

Color fixation was the first specific effect of nitrite recognized (National Academy of Sciences, 1981), and the minimal sodium nitrite concentration necessary for color fixation (∿40-50 mg/kg) is often the starting point for devising combinations of agents to duplicate the spectrum of effects currently obtained with higher sodium nitrite concentrations (120-200 mg/kg). Nevertheless, very little is known about some of the questions mentioned above. The information that is available on nitrite and possible alternatives is described below.

CONSUMER PREFERENCES FOR CURED-MEAT COLOR

Consumer color preferences in cured meats have not been extensively investigated. Preference may be displayed at the retail outlet, in some cases when the product is raw (as with bacon), or on consumption after cooking.

DuBose _et al_. (1981) evaluated the influence of ham color on flavor ratings. When color was concealed, low-nitrite samples were given higher ratings. When less red was perceived in some samples, a lower rating was given to cured flavor. This "color-cuing" effect on flavor occurred at concentrations of nitrite lower than that currently used to cure ham. However, consumers gave equal preference ratings to cooked nitrite-cured and nitrite-free bacon, even though the color difference was not concealed (Wasserman _et al_., 1977). A study of plate waste of bacon in a student dining hall produced a similar conclusion: equal amounts of nitrite-cured and nitrite-free bacon were left uneaten (Williams and Greene, 1979).

As far as the Committee is aware, there have been no attempts to determine the strength of the putative consumer preference for the traditional color of cured products. Experiences in Norway during the recent reduction in the use of nitrite could be evaluated to shed some light on this issue (Høyem, 1977), but societal differences need to be borne in mind in extrapolating conclusions to the U.S. situation.

USE OF NITRITE FOR COLOR FIXATION

Chemical Aspects of the Formation of Cured-Meat Color by Nitrite

The color of both fresh and cured-meat products is attributable primarily to the hemoprotein pigment, myoglobin (Clydesdale and Francis, 1971; Fox, 1966; Giddings, 1974, 1977a,b; Govindarajan, 1973). Figure 7-1 portrays the dynamic equilibrium between the various forms of myoglobin and demonstrates that the color of a meat product depends on the oxidation state of the heme iron in the pigment and on the type of functional group of the sixth ligand of the iron (Fox, 1966). The color of raw or fresh muscle tissue, such as beef or pork, is due to the purplish-red pigment, myoglobin (Mb); the bright-red pigment, oxymyoglobin (O_2Mb); and the brown pigment, metmyoglobin (MMb) (Clydesdale and Francis, 1971; Reith and Szakály, 1967a). Many factors influence the stability of these pigments (Clydesdale and Francis, 1971; Fox, 1966; Giddings, 1977a,b), and they are not stable when muscle tissue is heated (Reith and Szakály, 1967a). To obtain a more stable red pigment in heated commercial meat products, nitrite is added before heating. Biochemical reactions

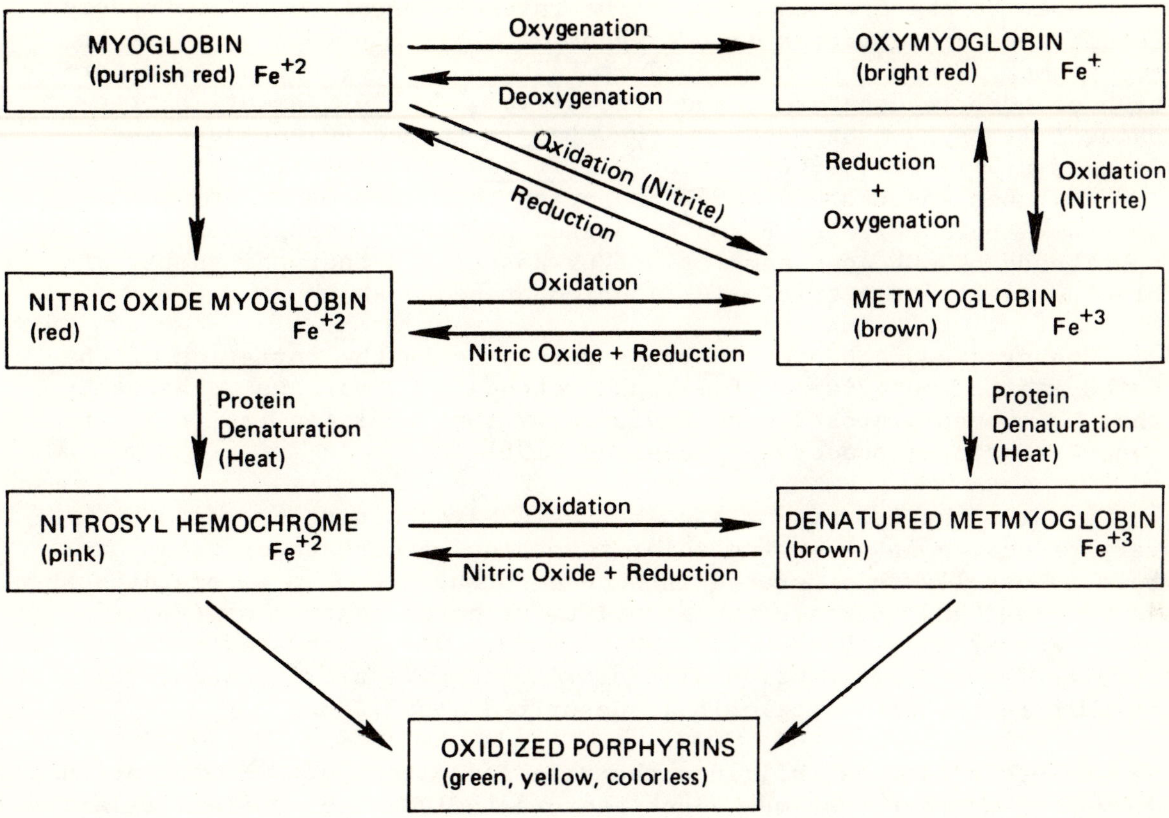

FIGURE 7-1. Some of the possible curing reactions that result from the addition of nitrite (Bard and Townsend, 1971).

in the meat reduce the nitrite to nitric oxide and the heme iron in myoglobin to the ferrous state. The interaction of these two species results in the formation of nitric oxide myoglobin (NOMb), a bright red pigment. When the meat product is then heated, the protein portion of NOMb is denatured, and a relatively stable pigment, nitric oxide (nitrosyl) myohemochrome (DNOMb), is formed (Clydesdale and Francis, 1971; Fox, 1966; MacDougall _et al_., 1975; Reith and Szakály, 1967a). Lee and Cassens (1976) reported that the amount of nitrogen-15 from labeled nitrite bound by heated samples of myoglobin was twice that bound by unheated samples. They postulated that the number of binding sites for nitric oxide doubled on heating.

The precise sequence of events resulting in the formation of the cured-meat pigment is not fully understood. The various mechanisms that have been proposed were reviewed by the Committee in its first report (National Academy of Sciences, 1981).

Because the isomers of ascorbic acid have been reported to be very effective in promoting color formation, ascorbate or isoascorbate is usually incorporated into cure formulations to accelerate the development and increase the stability of cured-meat color (Fox, 1966; Fox _et al_., 1967; Watts and Lehmann, 1952). This effect results from their reaction with nitrite to give nitric oxide, which in turn reacts with myoglobin as described above.

Numerous factors influence the rate and extent of NOMb formation in model and meat systems. Such factors include the type and relative concentrations of exogenous reductants (such as ascorbate), pH, the presence of salt and metal ions, storage temperature, amount of nitrite added, temperature reached during heating or cooking, exclusion of oxygen during formulation, and the original pigment concentration in the meat (Acton _et al_., 1979; Fox _et al_., 1967; Giddey, 1966; Reith and Szakály, 1967a,b; Renerre and Rougie, 1979; Siedler and Schweigert, 1959; Watts and Lehmann, 1952; Weiss _et al_., 1953).

Once formed, the complex of myoglobin and nitric oxide is very stable in the absence of oxygen. In the presence of oxygen, which rapidly oxidizes free nitric oxide to nitrite, the stability of the complex is limited by the rate of nitric oxide dissociation, because oxygen does not react directly with the bound nitric oxide (Giddings, 1977a,b). The dissociation is generally slow, occurring by autoxidation in air (Walsh and Rose, 1956), oxidation by nitrous acid (Walsh and Rose, 1956), lipid peroxide-induced oxidation (Younathan and Watts, 1959), or photocatalyzed oxidation (Bailey _et al_., 1964; Walsh and Rose, 1956). The underlying principle for all such mechanisms of nitric oxide-heme dissociation is believed to involve the withdrawal of electron density from iron to porphyrin, which weakens the Fe-NO bond. The nitric oxide group dissociates, leaving the iron susceptible to oxidation by the electronegative groups present in the medium (Tarladgis, 1962a,b). Such color loss is believed to be delayed by the stronger reducing conditions in the medium (Bailey _et al_., 1964; Lin _et al_., 1980; Reith and Szakály, 1967a; Tarladgis, 1962a,b), by the incorporation of nitrite at concentrations exceeding that of Mb (Reith and Szakály, 1967a,b; Tarladgis, 1962a,b; Walsh and Rose, 1956), by prevention of exposure to energy-generating electronic excitation (e.g., light) (Tarladgis, 1962a,b; Walsh and Rose, 1956), by replacement of NO-based curing salts with nitrogenous compounds that have strong electron-donor capability (Siedler and Schweigert, 1959; Tarladgis, 1962a,b), by elimination of oxygen during storage (Fox, 1966; Reith and Szakály, 1967b), by use of packaging films with low oxygen permeability (7 cm^3/m^2 per 24 h per bar) combined with a maximal extent of initial vacuum (687-737 mm Hg) (Kraft and Ayres, 1954; Lin and Sebranek, 1979; Lin *et al*., 1980), or by an increase in the pH of the product (Bailey *et al*., 1964; Reith and Szakály, 1967a; Walsh and Rose, 1956).

Various studies have been conducted to determine the minimal added nitrite necessary to produce cured-meat color in various products. In view of the many factors that affect color formation and stability, it is not surprising that there is some disagreement among studies.

Sensory Evaluation of Nitrite Concentration Needed for Cured-Meat Color

Only a small fraction of the nitrite added to a meat product is used for color fixation. Theoretically, only 3 mg of sodium nitrite per kilogram of product should provide a 50% conversion of Mb to NOMb (MacDougall *et al*., 1975). However, more is usually necessary to provide color stability, because of the effects of the many factors mentioned above, which influence the stability of the nitrosyl hemoprotein pigments, and because of the reaction of nitrite with other meat components, such as sulfhydryl and amino groups (Cassens *et al*., 1974, 1979; Woolford and Cassens, 1977).

Kerr *et al*. (1926) noted that incomplete color formation resulted from insufficient nitrite penetration into the meat or from unusually low myoglobin concentrations. This is exemplified by the fact that the minimal nitrite necessary to produce the desired color varies with the type of meat product, method of preparation, and presence of reductants, such as ascorbate (MacDougall *et al*., 1975; Sofos *et al*., 1979).

Using hedonic (preference) scales, Kemp *et al*. (1974, 1975), Eakes and Blumer (1975), and Eakes *et al*. (1975) found that the application of sodium nitrite (at $\sim$ 250 mg/kg), potassium nitrate (at up to 3,300 mg/kg), or their combination, to dry-cured hams resulted in color that was ranked more desirable (darker red) than the brownish-gray observed in the salt- and sugar-treated (control) sample. In dry-cured hams and pork loins cured with sodium nitrite, potassium nitrate, or both at 70-160 mg/kg, color was ranked significantly more acceptable than that of products "cured" with no nitrate or nitrite (Eakes and Blumer, 1975). DuBose *et al*. (1981) reported that pickle-cured, smoked hams (prepared with sodium nitrite at 25, 75, or 156 mg/kg) were not significantly different from each other in color, but their color was ranked significantly more acceptable than that of nitrite-free hams. However, Brown et al. (1974) reported a difference in color intensity between pickle-cured hams prepared with sodium nitrite at 91 and at 182 mg/kg; the higher concentration yielded darker color. The color of smoked turkey drumsticks cured in a pickling solution containing nitrite was found to be more acceptable than the color of their nitrite-free counterparts (Olson et al., 1979).

In comparison with pickle- or dry-cured products, comminuted meats require less nitrite for color development, because the chopping-emulsification process increases the available surface area and improves the distribution of nitrite. Wasserman and Talley (1972) and Hustad et al. (1973) reported gray color in unsmoked frankfurters prepared without nitrite in the cure. Similar results were found in the sensory evaluation of salami sausage (Skjelkvale et al., 1974) and Thüringer sausage (Dethmers et al., 1975). Hustad et al. (1973) reported no significant difference in color among frankfurters prepared with sodium nitrite concentrations of 50, 100, and 156 mg/kg. The lowest concentration of nitrite used with smoking imparted a characteristic color. Concentrations of sodium nitrite as low as 40 mg/kg resulted in acceptable color in chicken frankfurters (J. I. Gray, Michigan State University, Lansing, personal communication, 1981), in bacon (Paquette et al., 1980), and in turkey frankfurters (Sales et al., 1980), whereas 50 mg/kg was necessary for characteristic color to appear in a beef-pork bologna product (Lin and Sebranek, 1979) and in Thüringer sausage (Dethmers et al., 1975). In general, as the amount of added nitrite is increased, color acceptability of products increases and the products' color, as indicated by Hunter color values, becomes more red and less yellow (Sales et al., 1980; Sebranek et al., 1977).

In general, the use of low concentrations of sodium nitrite (∿50 mg/kg) can provide the characteristic pink color of cured meats. The use of alternative antimicrobial agents, such as sorbate, or additional processing, such as irradiation, does not appear to interfere with the coloring properties of nitrite. For example, Wierbicki and colleagues sought the lowest nitrite concentrations that would promote development of acceptable color in a variety of irradiated meats. The concentrations were: bacon, 20 mg/kg; corned beef and frankfurters, 50 mg/kg; and ham, 25 mg/kg. However, ham also required the addition of nitrate (25 mg/kg) to achieve stable color (Wierbicki and Brynjolfsson, 1979).

TESTING OF COMPOUNDS FOR PRODUCTION OF CURED-MEAT COLOR

There are relatively few reports of attempts to find compounds or processes that mimic the selective color "fixing" effect of nitrite in the muscle tissue of cured meats. Various patents have been filed on such compounds and processes (Table 7-1), but none has been adopted by the industry. Kemp (1974) cited the earlier work of

TABLE 7-1

Patents Granted for Meat Coloring Effects

Agent or Process	Patent
Nicotinic acid	Can. 980,579
Imidazole	Br. 1,173,446
Ascorbic acid plus one of: *p*-aminobenzoic acid (PABA) *m*-aminobenzoic acid (MABA) isonicotinic acid *N*-ethyl nicotinamide	U.S. 3,597,236
Tetrazole (1,2,3,5-tetrazole-NHN:NCH:N), aqueous at pH 4.5-6.3	Br. 1,251,644 Can. 886,515
Mixture of phosphates and sulfate, sulfite, or hyposulfite	Ja. 10986/72
Nitric oxide plus an ene-diol or a diketone reducing agent in presence of iron or iron salt	U.S. 37,780,192
Creation of reducing environment causing myoglobin-to-myochrome conversion, which could be complemented with nonmuscle pigments (e.g., betamin, erythrosine-red No. 3B, or rhodamine B)	Roczniki Institututu Presmyslu Miesnego 9(2)43-77(1972)
Carboxyhemoglobin (porcine) addition to product	Br. 1,294,415 (1972)

TABLE 7-2

Some Compounds Identified as Potential Substitutes for Nitrite in Cured-Meat Color Production[a]

Nicotinic Acid	Imidazole
Nicotinamide	Pyrazine
Hexylnicotinate	Triazine
Methylnicotinate	Nicotinic acid hydroxamate
Pyridine	2-Methylimidazole
Tetrazole	*N*,*N*,-dimethyl-nicotinamide
Purines	Trigonelline
Pyrimidines	Pentaerythritol
Pentaerythritol-tetranicotinate	

[a]From Kemp, 1974.

Brown's group, which had partial success in duplicating cured-meat color with the compounds listed in Table 7-2. Dymicky *et al*. (1975) cited other early studies on the problem.

The most extensive and systematic searches for color-producing agents have been conducted by Dymicky *et al*. (1975) and under the auspices of the Oscar Mayer Company, Madison, Wisconsin (P. Roehrig, 1980, personal communication). Both groups used a heat-processed meat system and tested a wide variety of compounds. In the Oscar Mayer study (described more fully in Table 7-3), 18 of 639 chemicals tested formed a pink color in pork under the experimental conditions used. These compounds are listed in Table 7-3.

In the study by Dymicky *et al*. (1975), frankfurter or a diluted frankfurter emulsion slurry was used, and selected compounds from a variety of chemical classes were tested for color-producing ability after heat processing at 70°C. Those most effective in producing the desired stable pink color were pyridine derivatives or related compounds, for example, 3-acylpyridines, quinoline, pyrazine, and imidazole.

Kanner and Juven (1980) reported that nitrite and *S*-nitrosocysteine, a compound generated from nitrite during curing, produced similar pink color in turkey meat when used at concentrations of 0.36, 0.73, and 1.45 mM (1 mM *S*-nitrosocysteine = 187 mg/kg; 1 mM sodium nitrite = 69 mg/kg). The anticlostridial effects of this compound are discussed in Chapter 4.

TABLE 7-3

Compounds That Produce a Pink Color in Heated Pork[a]

Carbon monoxide
Indazole
Cupferron
Amylnicotinate
3-Benzoylpyridine
3-Bromopyridine
Butylnicotinate
N,*N*-Diethylnicotinamide
4-Dimethylaminopyridine
Di-2-pyridyl ketone oxime
N-Ethylnicotinamide
Ethylnicotinate
Hexylnicotinate
Isonicotinate
Methylnicotinate
n-Propylnicotinate
Pyrazine
2-Pyridinealdoxime

---

[a]Study conducted by the Oscar Mayer Company, Madison, Wisconsin, under the following experimental conditions: Fresh postrigor pork was collected from production and ground twice through a 3-mm-pore-diameter plate. Sodium chloride and sodium ascorbate were added to the ground meat to achieve final concentrations of 20,000 and 550 mg/kg, respectively; 1 g and 0.1 g of the chemical to be evaluated were added to 100-g samples of meat. After mixing, the meat was placed in 38 x 300-mm test tubes and heated in a water bath at 82°C for 3 h. Samples were evaluated by visual observation at 0.5-h intervals.

Information on nonabsorbable food coloring agents was supplied to the Committee (Furia, 1977; N. M. Weinshenker, 1981, Dynapol, Palo Alto, California, personal communication). These agents and others, such as food dyes and naturally occurring pigments (e.g., betalines), impart a generalized color to products and may have application in finely comminuted (emulsified) or other products in which selective muscle-tissue color fixation is not essential.

Smoking has long been used as a method of preserving and imparting surface color and flavor to a variety of cured and uncured products (Hollenbeck, 1977). Generation of "smoke" may simultaneously produce nitrogen oxides (including nitric oxide), which are available for involvement in reactions that produce the nitrosyl hemopigments, as do nitrite reactions. Smoking may thus be regarded as an alternative means of producing cured-meat color, but the mechanism by which this occurs and the toxicologic implications have not been thoroughly evaluated.

CONCLUSIONS AND RECOMMENDATIONS

The strength of the putative consumer preference for the traditional pink-red color in cured products has not been adequately investigated. Although consumer preference could probably be modified, to what degree is unknown.

Investigation of color discriminability in various cured products by analytic sensory panels and acceptance-preference tests by representative consumers should be accorded high priority. Recent experiences in Norway with reduction in nitrite addition might usefully be examined in this regard, but societal differences need to be borne in mind in extrapolating conclusions to the U.S. situation.

The data suggest that an adequately stable uniform cured-meat color can be achieved in most cured products with sodium nitrite added at approximately 50 mg/kg. Other than the use of reduced concentrations of nitrite, no suitable means of fixing color in cured meats has been demonstrated to be effective in products made under commercial conditions.

Research directed to finding alternatives to nitrite for color fixation should focus on agents that could confer color selectively to the muscle tissue of cured meats. Investigation of the potential of food colors and nonabsorbable coloring agents to replace nitrite in some products, such as frankfurters, may be warranted, but it should not be accorded high priority.

The compounds shown to be effective in producing cured-meat color in model meat systems should be evaluated in meat products by chemical and sensory testing. If satisfactory results are obtained, their toxic characteristics should be evaluated as outlined in Chapter 10 before they are tested on a commercial scale.

The Committee recommends that studies be conducted to determine the extent to which nitrogen oxides from "smoke" contribute to color in products that do not contain nitrite.

REFERENCES

Acton, J. C., R. L. Dick, and A. K. Torrence. 1979. Turkey ham properties on processing and cured color formation. Poult. Sci. 58:843-847.

Bailey, M. E., R. W. Frame, and H. D. Naumann. 1964. Cured meat pigments: Studies of the photooxidation of nitrosomyoglobin. J. Agric. Food Chem. 12:89-93.

Bard, J., and W. E. Townsend. 1971. Meat curing. Pp. 452-483 in J. F. Price and B. S. Schweigert, eds. The Science of Meat and Meat Products. J. H. Freeman and Company, San Francisco.

Brown, C. L., H. B. Hedrick, and M. E. Bailey. 1974. Characteristics of cured ham as influenced by levels of sodium nitrite and sodium ascorbate. J. Food Sci. 39:977-979.

Cassens, R. G., J. G. Sebranek, G. Kubberod, and G. Woolford. 1974. Where does the nitrite go? Food Prod. Dev. 8:50-56.

Cassens, R. G., M. L. Greaser, T. Ito, and M. Lee. 1979. Reactions of nitrite in meat. Food Technol. 33:46-57.

Clydesdale, F. M., and F. J. Francis. 1971. Color measurement of foods. XXVII. Chemistry of meat color. Food Prod. Dev. 5:81-82, 87-90.

Dethmers, A. E., H. Rock, T. Fazio, and R. W. Johnston. 1975. Effect of added sodium nitrite and sodium nitrate on sensory quality and nitrosamine formation in thuringer sausage. J. Food Sci. 40:491-495.

DuBose, C. N., A. V. Cardello, and O. Maller. 1981. Factors affecting the acceptability of low-nitrite smoked, cured ham. J. Food Sci. 46:461-463.

Dymicky, M., J. B. Fox, and A. E. Wasserman. 1975. Color formation in cooked model and meat systems with organic and inorganic compounds. J. Food Sci. 40:306-309.

Eakes, B. D., and T. N. Blumer. 1975. Effect of various levels of potassium nitrate and sodium nitrite on color and flavor of cured loins and country-style hams. J. Food Sci. 40:977-980.

Eakes, B. D., T. N. Blumer, and R. J. Monroe. 1975. Effect of nitrate and nitrite on color and flavor of country-style hams. J. Food Sci. 40:973-976.

Fox, J. B., Jr. 1966. The chemistry of meat pigments. J. Agric. Food Chem. 14:207-210.

Fox, J. B., W. E. Townsend, S. A. Ackerman, and C. E. Swift. 1967. Cured color development during frankfurter processing. Food Technol. 21:386-392.

Furia, T. E. 1977. Nonabsorbable, polymeric food colors. Food Technol. (Chicago) 31:34-37, 39.

Giddey, C. 1966. The change in meat pigments in sausage making processes. J. Sci. Food Agric. 17:14-17.

Giddings, G. G. 1974. Reduction of ferrimyoglobin in meat. Crit. Rev. Food Technol. 5:143-173.

Giddings, G. G. 1977a. The basis of color in muscle foods. Crit. Rev. Food Sci. Nutr. 8:81-114.

Giddings, G. G. 1977b. Symposium: The basis of quality in muscle foods. The basis of color in muscle foods. J. Food Sci. 42:288-294.

Govindarajan, S. 1973. Fresh meat color. CRC Crit. Rev. Food Technol. 4:117-140.

Hollenbeck, C. M. 1977. Novel concepts in technology and design of machinery for production and application of smoke in the food industry. Pure and Appl. Chem. 49:1687-1702.

Høyem, T. 1977. Experiences with nitrite and nitrate ban in Norway. Pp. 293-297 in B. J. Tinbergen and B. Krol, eds. Proceedings of the Second International Symposium on Nitrite in Meat Products, Zeist, 1976, PUDOC, Wageningen, the Netherlands.

Hustad, G. O., J. G. Cerveny, H. Trenk, R. H. Deibel, D. A. Kautter, T. Fazio, R. W. Johnston, and O. E. Kolari. 1973. Effect of sodium nitrite and sodium nitrate on botulinal toxin production and nitrosamine formation in wieners. Appl. Microbiol. 26:22-26.

Jeremiah, L. E., Z. L. Carpenter, and G. C. Smith. 1972. Beef color as related to consumer acceptance and palatability. J. Food Sci. 37:476-479.

Kanner, J., and B. J. Juven. 1980. S-Nitrosocysteine as an antioxidant, color-developing, and anticlostridial agent in comminuted turkey meat. J. Food Sci. 45:1105-1112.

Kemp, J. D. 1974. Nitrate and nitrite substitutes in meat curing. Food Prod. Dev. 8:64, 67-68, 70.

Kemp, J. D., J. D. Fox, and W. G. Moody. 1974. Cured ham properties as affected by nitrate and nitrite and fresh pork quality. J. Food Sci. 39:972-976.

Kemp, J. D., B. E. Langlois, J. D. Fox, and W. Y. Varney. 1975. Effects of curing ingredients and holding times and temperatures on organoleptic and microbiological properties of dry-cured sliced ham. J. Food Sci. 40:634-636.

Kerr, R. H., C. T. N. Marsh, W. F. Schroeder, and E. A. Boyer. 1926. The use of sodium nitrite in the curing of meat. J. Agric. Res. 33:541-551.

Kraft, A. A., and J. C. Ayres. 1954. Color changes in packaged bologna. Food Technol. 8:22-26.

Lee, S. H., and R. G. Cassens. 1976. Nitrite binding sites on myoglobin. J. Food Sci. 41:969-970.

Lin, H.-S., and J. G. Sebranek. 1979. Effect of sodium nitrite concentration and packaging conditions on color stability and rancidity development in sliced bologna. J. Food Sci. 44:1451-1454.

Lin, H.-S., J. G. Sebranek, D. E. Galloway, and K. D. Lind. 1980. Effect of sodium erythorbate and packaging conditions on color stability of sliced bologna. J. Food Sci. 45:115-118.

MacDougall, D. B., D. S. Mottram, and D. N. Rhodes. 1975. Contribution of nitrite and nitrate to the colour and flavour of cured meats. J. Sci. Food Agric. 26:1743-1754.

National Academy of Sciences. 1981. The Health Effects of Nitrate, Nitrite, and *N*-Nitroso Compounds. National Academy Press, Washington, D.C. 544 pp.

Olson, V. M., N. A. King, J. A. Langbehn, and W. J. Stadelman. 1979. Acceptability of smoked turkey drumsticks with and without nitrite addition. Poult. Sci. 58:587-590.

Paquette, M. W., M. C. Robach, J. N. Sofos, and F. F. Busta. 1980. Effects of various concentrations of sodium nitrite and potassium sorbate on color and sensory qualities of commercially prepared bacon. J. Food Sci. 45:1293-1296.

Reith, J. F., and M. Szakály. 1967a. Formation and stability of nitric oxide myoglobin. I. Studies with model systems. J. Food Sci. 32:188-193.

Reith, J. F., and M. Szakály. 1967b. Formation and stability of nitric oxide myoglobin. II. Studies on meat. J. Food Sci. 32:194-196.

Renerre, M., and P. Rougie. 1979. [In French; English Summary.] [Influence of heating on the nitrite binding on myoglobin.] Ann. Technol. Agric. 28:423-431.

Sales, C. A., J. A. Bowers, and D. Kropf. 1980. Consumer acceptability of turkey frankfurters with 0, 40, and 100 ppm nitrite. J. Food Sci. 45:1060-1061.

Sebranek, J. G., B. G. Schroder, R. E. Rust, and D. G. Topel. 1977. Influence of sodium erythorbate on color development, flavor and overall acceptability of frankfurters cured with reduced levels of sodium nitrite. J. Food Sci. 42:1120-1121.

Siedler, A. J., and B. S. Schweigert. 1959. Biochemistry of myoglobin: Effects of heat, nitrite level, iron salts, and reducing agents on formation of denatured nitrosomyoglobin. J. Agric. Food Chem. 7:271-274.

Skjelkvale, R., T. B. Tjaberg, and M. Valland. 1974. Comparison of salami sausages produced with and without addition of sodium nitrite and sodium nitrate. J. Food Sci. 39:520-524.

Sofos, J. N., F. F. Busta, and C. E. Allen. 1979. Botulism control by nitrite and sorbate in cured meats: A review. J. Food Prot. 42:739-770.

Tarladgis, B. G. 1962a. Interpretation of the spectra of meat pigments. I. Cooked meats. J. Sci. Food Agric. 13:481-484.

Tarladgis, B. G. 1962b. Interpretation of the spectra of meat pigments. II. Cured meats. The mechanism of colour fading. J. Sci. Food Agric. 13:485-491.

Walsh, K. A., and D. Rose. 1956. Meat pigments: Factors affecting the oxidation of nitric oxide myoglobin. J. Agric. Food Chem. 4:352-355.

Wasserman, A. E., and F. Talley. 1972. The effect of sodium nitrite on the flavor of frankfurters. J. Food Sci. 37:536-538.

Wasserman, A. E., W. Kimoto, and J. G. Phillips. 1977. Consumer acceptance of nitrite-free bacon. J. Food Prot. 40:683-685.

Watts, B. M., and B. T. Lehmann. 1952. The effect of ascorbic acid on the oxidation of hemoglobin and the formation of nitric oxide hemoglobin. Food Res. 17:100-108.

Weiss, T. J., R. Green, and B. M. Watts. 1953. Effect of metal ions on the formation of nitric oxide hemoglobin. Food Res. 18:11-16.

Wierbicki, E., and A. Brynjolfsson. 1979. The use of irradiation to reduce or eliminate nitrite in cured meats (paper 8.2). Presented at the 25th European Meeting of Meat Research Workers, August 27-31, 1979, Budapest, Hungary.

Williams, J. C., and B. E. Greene. 1979. Plate waste of bacon cured with and without sodium nitrite. J. Food Sci. 44:1260, 1262.

Woolford, G., and R. G. Cassens. 1977. The fate of sodium nitrite in bacon. J. Food Sci. 42:586-589, 596.

Younathan, M. T., and B. M. Watts. 1959. Relationship of meat pigments to lipid oxidation. Food Res. 24:728-734.

CHAPTER 8

METHODS TO REDUCE NITROSAMINE FORMATION IN CURED MEATS

CHAPTER 8

METHODS TO REDUCE NITROSAMINE FORMATION IN CURED MEATS

Under appropriate conditions, nitrite can react with nitrosatable amino compounds to form *N*-nitroso compounds. Nitrosation reactions can occur in the environment or in the body. Because of the ability of most nitrosamines that have been tested thus far to induce cancer in laboratory animals, exposure to these chemicals is considered to present a potential cancer risk to humans. Thus, it is desirable to reduce human exposure to these compounds.

In its first report, the Committee estimated that the average U.S. nonsmoker is exposed to approximately 3 μg of *N*-nitroso compounds per day. Of this amount, ∿1.7 μg comes from exogenous sources (e.g., food, cosmetics, and alcoholic beverages) and ∿1.3 μg is formed endogenously. Thus, the most effective means of reducing risks associated with exposure to *N*-nitroso compounds would decrease their formation exogenously and endogenously. Unfortunately, little is known about factors influencing the endogenous synthesis of these compounds (National Academy of Sciences, 1981); thus, this chapter presents evidence only on methods that reduce exogenous formation of *N*-nitroso compounds. However, it is possible that methods developed to decrease nitrosation reactions *in vitro* may be effective *in vivo* as well.

The food items of major concern with respect to the exogenous formation of nitrosamines, a class of *N*-nitroso compounds, are cured meat products, especially bacon. Nitrosopyrrolidine (NPYR) and, to a lesser extent, nitrosodimethylamine (NDMA) are the two nitrosamines isolated most frequently from this product (Gray and Randall, 1979; Pensabene *et al*., 1974, 1980; Sen *et al*., 1979), where NPYR is found almost invariably in cooked, but not raw, bacon. The concentration of NPYR in cooked bacon depends on such factors as the method of cooking, frying temperature and time, nitrite concentration, ascorbate concentration, preprocessing procedures, lean-to-adipose tissue ratio, presence of lipophilic inhibitors, and, possibly, smoking (Gray and Randall, 1979).

Most of the studies on nitrosamine formation in meat have been conducted in pickle-cured bacon. Two recent investigations, however, have indicated the presence of high concentrations of NPYR in dry-cured bacon after frying. NPYR concentrations ranging from 39 to 89 μg/kg (Pensabene *et al*., 1979) and from traces to 320 μg/kg (Nitrite Safety Council, 1980) have been reported. Other dry-cured products in which nitrosamines have been detected include dry-cured hams and shoulders (Nitrite Safety Council, 1980). As was the case with bacon, nitrosamines were detected only in samples that had been fried. Studies of other cured meat products have indicated that, although nitrosamines may be detected, they are usually present at very low concentrations (< 1 μg/kg) (Gough *et al.*, 1978; Nitrite Safety Council, 1980; Sen *et al.*, 1979).

Thus, the following discussion emphasizes studies that have shown a reduction in nitrosamine formation in bacon. However, because nitrite present in other cured meats may participate in nitrosation reactions *in vivo*, methods similar to those described for reducing the nitrosamine content of bacon may, in the future, be applied to other nitrite-containing products as well.

There are two general approaches to reducing nitrosamine formation in meat products to which nitrite is added. One is to eliminate or reduce nitrite in the product. (Virtually all the alternatives discussed in previous chapters of this report are being investigated primarily as a means to reduce the nitrite content of cured meats and thus reduce nitrosamine formation.) The impact of the reduction of nitrite concentrations on the formation of nitrosamines in cured meats is discussed in this chapter. The other approach is to add an inhibitor of nitrosation reactions to the product during its processing. The Committee has reviewed available data on the inhibition of nitrosamine formation in cured meats and other media and a summary of these data appears later in this chapter.

Although the patent literature contains several additional methods that have been proposed to reduce nitrosamine formation (Kelly, 1974), many of these--including proposals to replace the sodium nitrite with butylnitrite during the curing of bacon (Bharucha *et al.*, 1976) or to use a two-step curing process involving sodium nitrite in the first step and sodium metabisulfite in the second step (Coleman *et al.*, 1975)--have not been studied in depth and are not included in our discussion.

One other method of reducing nitrosamine formation in dry-cured bacon has been reported by Bailey (1980). He found that the addition of 1-2% dextrose to the curing mixture reduced the concentration of NPYR formed. This finding indicates that further study of the effect of cure formulation and other factors in the processing of dry-cured products on nitrosamine formation may suggest other possible methods of reducing the amounts of these compounds in such products.

EFFECT OF REDUCED NITRITE CONCENTRATIONS

The effects of various amounts of nitrite on the formation of NPYR during the cooking of bacon have been studied by Sen and Donaldson (1974), who used bacon samples prepared with nitrite at 0, 50, 100, 150, and 200 mg/kg. These authors demonstrated that the amounts of NPYR formed in fried bacon increased as the nitrite concentration increased. The amounts of NPYR produced correlated well with the initial concentration, but not the final (residual) concentration of nitrite (analyzed before frying). Herring (1973) also showed that increasing the nitrite concentration markedly increased NPYR formation (Table 8-1).

TABLE 8-1

Effect of Added Sodium Nitrite on Formation of Nitrosopyrrolidine (NPYR) During Cooking of Bacon[a]

Sample and Cooking Method	NPYR Concentration, μg/kg, After Addition of Sodium Nitrite, mg/kg, At:					
	0	15	30	60	120	140
Bacon						
Pan-fried	0	0	5	10	14	24
Oven-fried	0	0	0	6	6	14
Microwave-cooked	0	0	4	3	5	2
Fat cookout						
Pan-fried	0	0	10	10	26	30
Oven-fried	0	0	4	6	15	14
Microwave-cooked	0	0	4	5	12	16

[a]Table adapted from Herring (1973), with permission.

Although these studies suggested that the initial, and not the residual, nitrite concentration influences nitrosamine formation in bacon, recent evidence indicates that the lowest residual nitrite leads to the least nitrosamine formation (Dudley, 1979; Sebranek, 1979).

Because residual nitrite concentration depends on the initial concentration (as well as other factors, such as duration and temperature of storage), the U.S. Department of Agriculture (1978) recommended that the initial concentration of sodium nitrite in bacon be reduced from 156 to 120 mg/kg. Similarly, in Canada, the concentration of nitrite to be used in the preparation of side bacon has been reduced from 200 to 150 mg/kg--calculated before any smoking, cooking, or fermentation (Gray, 1976). Periodic surveillance over the last few years has suggested that the concentration of nitrosamines in bacon has been steadily decreasing in both the United States (Havery *et al*., 1978; Table 8-2) and Canada (Sen *et al*., 1977). The trend toward lower NPYR concentrations in cooked bacon is partially explained by the use of reduced amounts of nitrite and increased amounts of the nitrosation inhibitor, ascorbate, in the bacon-curing mixture (Havery *et al*., 1978).

TABLE 8-2

Concentrations of NPYR in Commercial U.S. Cooked Bacon[a]

	Concentration, μg/kg, by Brand[b]								
Year	A	B	C	D	E	F	G	H	I
1971	-	-	-	-	100	77	73	-	-
1972	110	-	20,13,24	100	-	-	95	-	25
1973	34	58,36	-	25	39	29	86,79	-	25
1974	17	-	14,8	7	10	29	96	45,13	38
1975	-	18	5	12	19	17	65	13,10	12
1976	18	25	-	3	23	7	33	6	21
1977	29	14	5	5	12	-	75	10	9

[a]Adapted from Havery *et al*., 1978.
[b]Nine brands of bacon purchased at various times in 1971-1977 in a Washington, D.C. retail market. Some concentrations have been rounded to two significant figures.

Combinations of nitrite at lower concentrations and an antimicrobial agent at concentrations sufficient to inhibit microbial proliferation have been investigated to determine their effect on nitrosamine formation. For example, Robach *et al*. (1980) investigated the effect of various concentrations of sodium nitrite and potassium sorbate on nitrosamine formation in commercially prepared bacon. Bacon processed with nitrite at 40 mg/kg and 0.26% sorbate contained NPYR at an average of 8.7 μg/kg, whereas samples prepared with nitrite at 120 mg/kg contained NPYR at an average of 28.1 μg/kg. This marked reduction in NPYR is probably due to the reduction in nitrite, although it has been reported that sorbic acid also has an inhibitory effect on nitrosation reactions (Tanaka *et al*., 1978).

EFFECT OF NITROSATION INHIBITORS

The formation of nitrosamines can be reduced or eliminated by the presence of chemicals that reduce dinitrogen trioxide (N_2O_3), the nitrosating species, to the nonnitrosating nitric oxide (NO). They include thiols, sulfur dioxide, bisulfite, ascorbic acid (vitamin C), α-tocopherol (vitamin E), gallic acid, and other synthetic and natural antioxidants. Data on the effectiveness of these agents *in vitro* and *in vivo* have been reviewed by Douglass *et al*. (1978).

The effectiveness of an inhibitor depends on the concentration of nitrite, the concentration of inhibitor (which must be greater than that of nitrite), the concentration of the nitrosatable amine, the ease of nitrosation of the amino substrate, and various reaction conditions, such as pH, temperature, and the medium in which the reaction takes place. In the case of nitrosation reactions taking place in cured meats (especially bacon), the following characteristics are thought to be essential for an effective inhibitor: solubility in lipids, nonvolatility in steam, and stability at temperatures up to 174°C (Bharucha *et al*., 1979). These conclusions are based on the observation that nitrosamine formation occurs primarily in the lipid phase during the frying of bacon. Of the nitrosation inhibitors listed above, several have been tested in cured meats (primarily bacon), including ascorbic acid, α-tocopherol, and other antioxidants.

Ascorbic Acid

Ascorbic acid has been shown to be an excellent inhibitor of nitrosation reactions, particularly in weakly acidic conditions (Mirvish *et al*., 1972). In the chemical reaction, nitrite is reduced to nitric oxide (NO), and ascorbic acid is oxidized to dehydroascorbic acid (Reaction 2 and Figure 8-1). The ascorbate anion also reacts with N_2O_3 (Mirvish, 1981; Reaction 3). Thus, amines (Reaction 4) and ascorbic acid compete directly for the nitrosating species (N_2O_3).

$$\underset{\text{(nitrous acid)}}{2HNO_2} \longrightarrow N_2O_3 + H_2O \qquad (1)$$

$$\underset{\text{(ascorbic acid)}}{C_6H_8O_6} + N_2O_3 \xrightarrow{k_1} \underset{\text{(dehydroascorbic acid)}}{C_6H_6O_6} + 2NO + H_2O \qquad (2)$$

$$\underset{\text{(ascorbate)}}{C_6H_7O_6} + N_2O_3 \xrightarrow{k_2} C_6H_6O_6 + 2NO + OH^- \qquad (3)$$

$$\underset{\text{(secondary amine)}}{R_2NH + N_2O_3} \xrightarrow{k_3} \underset{\text{(nitrosamine)}}{R_2NNO} + HNO_2 \qquad (4)$$

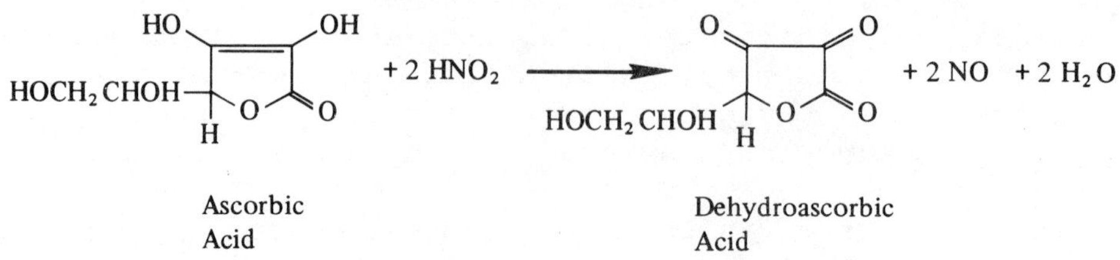

FIGURE 8-1. Reaction of ascorbic acid with nitrite.

According to Dahn *et al.* (1960), the ascorbate anion reacts with N_2O_3 approximately 230 times faster than ascorbic acid, presumably as a result of its greater nucleophilic activity. At a pH of 3-5, the anion predominates and its reaction with nitrite is so rapid that formation of N_2O_3 is rate-limiting.

The properties and utility of ascorbic acid as an inhibitor of nitrosamine formation have been well described in the literature (Archer *et al.*, 1975; Gray and Dugan, 1975; Kamm *et al.*, 1973, 1975, 1977; Mergens *et al.*, 1978; Mirvish, 1981; Newmark and Mergens, 1981). Ascorbic acid, sodium ascorbate, erythorbic acid, and sodium erythorbate have been added to cured meats for many years, primarily to improve their color characteristics (Bard and Townsend, 1971). After the initial report that ascorbate may reduce nitrosamine formation in nonfood systems (Mirvish *et al.*, 1972), Fiddler *et al.* (1973) showed that frankfurters prepared with ascorbic acid at 550 or 5,500 mg/kg and nitrite at 1,500 mg/kg and processed for 2 h had no NDMA present. Samples made with nitrite alone contained approximately 10 μg/kg of NDMA. Brown *et al.* (1974) reported that hams treated with sodium ascorbate had lower residual nitrite than non-ascorbate-treated hams, suggesting that the mechanism of nitrosamine inhibition was through the reaction of ascorbate with nitrite.

The effect of increasing the amount of ascorbic acid on NPYR formation in bacon has also been extensively studied (Herring, 1973). Bacon containing sodium ascorbate at 1,000 mg/kg did not produce any NPYR on frying. Decreasing the amount of ascorbate to 250 and 0 mg/kg increased the amounts of NPYR formed. Mottram *et al.* (1975) investigated the influence of ascorbic acid and pH on formation of NDMA in cured pork containing added dimethylamine (0.1%). Heating of the cured pork resulted in formation of small amounts of NDMA in the lean and considerably higher amounts in the fat portions of the meat. However, the lowest amounts of NDMA were found in the bacon containing ascorbate.

The recognition that ascorbate is an effective inhibitor of nitrosamine formation in meats has led to the regulation that all bacon be processed with ascorbate or erythorbate at 550 mg/kg (U.S. Department of Agriculture, 1978).

However, compounds of ascorbic acid are not completely successful in bacon as inhibitors, because of their low solubility in lipids, and studies of the effect of several ascorbates on nitrosamine formation in a model system resembling bacon fat revealed that ascorbate could increase nitrosamine formation under some conditions (Mottram, 1976; Mottram and Patterson, 1977). In these studies, the nitrosation of dipropylamine and pyrrolidine was examined in a two-phase system composed of an aqueous buffer and a nonpolar solvent. Contrary to previous results with purely aqueous systems, sodium ascorbate increased nitrosation by a factor of 5-25, compared with ascorbate-free controls. Lipophilic ascorbyl palmitate, however, reduced nitrosamine formation in this model system.

Ascorbyl palmitate has also been found (Bharucha *et al*., 1980; Sen *et al*., 1976) to be more effective than sodium ascorbate in reducing nitrosamine formation in fried bacon. Ascorbyl palmitate used at 500-1,000 mg/kg reduced nitrosamine formation in bacon by 70-90%, compared with samples without added inhibitor; however, this compound becomes less active with increased storage time (Bharucha *et al*., 1980).

The long-chain acetals of ascorbic acid (dodecanal, C_{12}; tetradecanal C_{14}; hexadecanal, C_{16}; octadecanal, C_{18}; and 9-octadecanal, C_{18}) streaked on bacon slices at 1,000 mg/kg have also been reported to be effective, bringing about a 93-98% reduction in nitrosamines present in the cooked-out fat (Bharucha *et al*., 1980). Unlike ascorbyl palmitate, the long-chain acetals retain their activities (90% inhibition of nitrosamine formation) for at least 35 d at $3^{o}C$ if applied to bacon at 1,000 mg/kg (Bharucha *et al*., 1980). These compounds have not been subjected to toxicologic testing and are not approved for use in food systems.

α-Tocopherol

α-Tocopherol, the major form of vitamin E, has been shown to be an effective nitrosation inhibitor in chemical systems (Kamm *et al*., 1977; Mergens *et al*., 1978); mechanisms of its activity are discussed in Chapter 4 of the Committee's first report. It also has the advantage over ascorbate of being soluble in lipids, and it is especially applicable for use in meat systems. As mentioned in the preceding section, lipids can act simultaneously as ready solvents for the unprotonated free-base amino substrate and the nitrous anhydride (N_2O_3). During the frying of bacon, residual nitrite present in the aqueous phase is dehydrated to yield N_2O_3 and is driven into the fat layers where nitrosation occurs.

α-Tocopherol functions as a blocking agent through its ability to reduce nitrite. During this reaction, the tocopherol is oxidized to a quinone (Figure 8-2). α-Tocopherol does not have unsubstituted carbon atoms in the phenolic ring, so it cannot form C-nitroso derivatives that might promote transnitrosation. However, β-, γ-, and δ-tocopherols have unsubstituted positions in the phenol ring and are not as effective as α-tocopherol in inhibiting nitrosation (Mirvish, 1981; Newmark and Mergens, 1981).

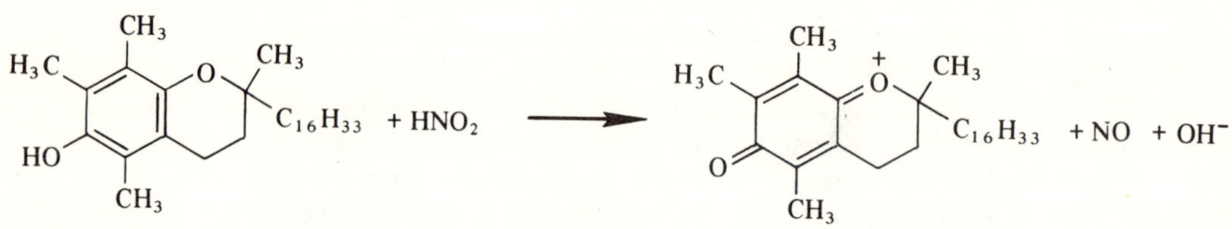

FIGURE 8-2. Reaction of α-tocopherol with nitrite.

Evidence that α-tocopherol is effective in meat comes from studies of the inhibition of NPYR formation in fried bacon (Fiddler *et al.*, 1978). A significant reduction of nitrosamine concentration was detected when α-tocopherol was used at 250-500 mg/kg.

Because α-tocopherol is not soluble in the aqueous cure mixture, special methods to increase solubility are required if the agent is to be applied as part of the cure. Fiddler *et al.* (1978) used polysorbate emulsifiers in the cure to obtain adequate distribution of α-tocopherol in the product. α-Tocopherol can also be effectively dispersed throughout the product during the frying of bacon; therefore, it may be applied to bacon by spraying or dipping (Mergens and Newmark, 1979) or by frying in fat containing α-tocopherol (Walters *et al.*, 1976).

Reddy and Gray (personal communication, 1981) successfully reduced NPYR formation in dry-cured bacon with α-tocopherol-treated salt systems. Preliminary results indicated an average inhibition of 96% of NPYR formation in pork bellies cured with the addition of α-tocopherol at 500 mg/kg.

α-Tocopherol, added in sufficient concentrations to inhibit nitrosamine formation to bacon containing ascorbate at 550 mg/kg, apparently does not interfere with the antibotulinal activity of nitrite in tests conducted shortly (1-10 d) after processing (Reddy and Gray, personal communication, 1981; Tanaka, 1980).

Other Agents

Polyphenolic compounds, such as gallic acid, and simple phenols can function as inhibitors of nitrosamine formation under some conditions. The inhibition is due to the interaction of phenols and nitrite, which results in either the formation of *C*-nitrosophenols or, in the case of polyphenols, the reduction of nitrite to nitric oxide. In the latter reaction, the phenols are oxidized to quinones (Challis, 1973; Mirvish, 1981). Pignatelli *et al*. (1980) have shown that 1,2- and 1,4-dihydroxyphenols (including naturally occurring flavonols) inhibit nitrosamine formation at a pH of 4.0 (Figure 8-3). However, 1,3-dihydroxyphenols (e.g., resorcinol) are powerful catalysts under similar conditions (Pignatelli *et al*., 1980). The catalysis by 1,3-dihydroxyphenols is probably due to the rapid formation of a nitroso intermediate that interacts with more dinitrogen trioxide to generate a powerful nitrosating agent (Figure 8-3).

Several sulfur compounds can also function as inhibitors. Sodium bisulfite has been extensively studied, and its effectiveness at a low pH has been found to equal that of ascorbic acid and α-tocopherol (Mirvish, 1975). At a pH of 1-4, bisulfite reduces nitrite in two steps (Hisatsune, 1961)--first to nitric oxide (Reaction 5) and then to nitrous acid (Reaction 6).

$$SO_2 + 2HNO_2 \longrightarrow 2NO + H_2SO_4 \qquad (5)$$

$$SO_2 + 2NO + H_2O \longrightarrow N_2O + H_2SO_4 \qquad (6)$$

Thiols, such as cysteine and glutathione, that are present in meat can reduce acidified nitrite to nitric oxide through the formation of nitrite thioesters. These interactions may inhibit the formation of *N*-nitroso compounds. However, thionitrite esters are stable above a pH of 7 and in lipophilic media (Davies *et al*., 1978), and they may promote, rather than inhibit, the formation of nitrosamines under such conditions.

A number of antioxidants have been tested in cured meats or model systems for their ability to inhibit nitrosation reactions, including

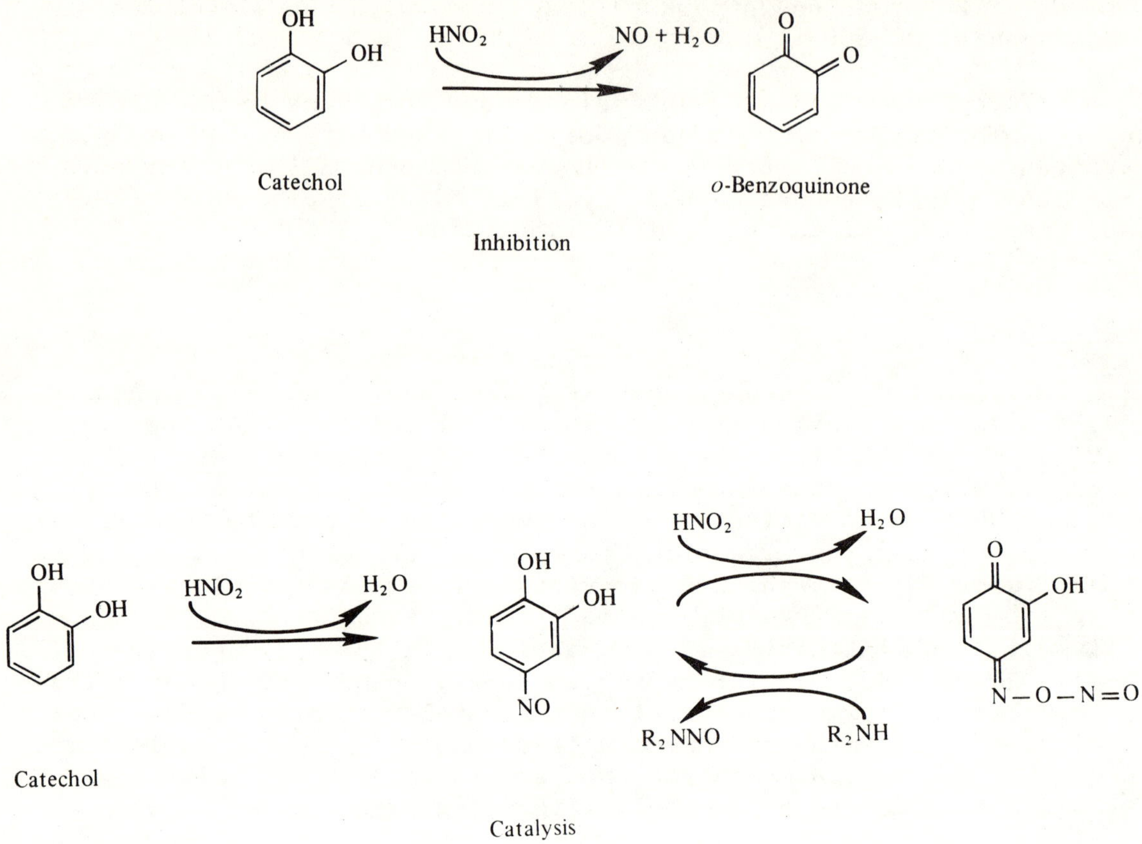

FIGURE 8-3. Mechanism whereby phenols can either inhibit (top) or increase (bottom) nitrosation (Mirvish, 1981). For simplicity, both reactions are shown for the same phenol, catechol. Reprinted by permission from Mirvish, 1981. Inhibition of formation of carcinogenic N-nitroso compounds by ascorbic acid and other compounds. Pp. 557-587 in J. H. Burchenal and H. F. Oettgen, eds. Cancer: Achievements, Challenges, and Prospects for the 1980s. Grune and Stratton, New York.

propyl gallate, ethoxyquin, butylated hydroxyanisole (BHA), and butylated hydroxytoluene (BHT). Sen *et al.* (1976) reported decreased nitrosamine concentrations in bacon that had been treated with propyl gallate immediately before frying. Ethoxyquin was shown to inhibit NPYR formation in a model system composed of proline-nitrite mixtures in methanol (Coleman, 1978). Additional studies of these agents in a simulated gastric environment revealed a lack of correlation between general antioxidant efficacy and ability to react with nitrite (i.e., inhibit nitrosation). For example, BHA and BHT, both effective lipid antioxidants, are very poor nitrosation inhibitors (Mergens and Newmark, 1980).

A summary of the activities of various nitrosation inhibitors in fried bacon is given in Table 8-3.

TABLE 8-3

Effects of Various Agents on Nitrosamine Formation in Fried Bacon

Compound	Concentration, mg/kg	Inhibition, % Fried Bacon	Inhibition, % Cooked-Out Fat	Reference
Ascorbic acid	1,000	100	--	Greenberg, 1974
Ascorbyl palmitate	1,000	59-87[a]	--	Sen *et al.*, 1976
	500-1,000	--	70-90	Bharucha *et al.*, 1980
Ascorbic acid acetal	1,000	62-88	90-98	Bharucha *et al.*, 1980
Propyl gallate	1,000	50-97[a]	--	Sen *et al.*, 1976
α-Tocopherol	500	80	--	Fiddler *et al.*, 1978
	500	85	--	Mergens and Newmark, 1979

[a]Analyses conducted on combined bacon and cooked-out fat extracts.

CONCLUSIONS AND RECOMMENDATIONS

Two general approaches exist for reducing the concentrations of nitrosamines in cured meats, particularly bacon. One method that has been proposed is to reduce nitrite addition to approximately 50 mg/kg (for color development) and include effective concentrations of an antimicrobial agent, such as sorbate. Another approach is to incorporate nitrosation inhibitors, such as ascorbic acid and α-tocopherol, into cured-meat products that contain the usual concentration of sodium nitrite.

Both approaches appear to be effective in reducing nitrosamine formation in cured meats, especially bacon. Reduced concentrations of nitrite in combination with the antimicrobial agent, sorbate, have been shown to be effective in preventing botulinal toxin production (see Chapter 4), as well as in reducing the concentration of nitrosamines.

Of the nitrosation inhibitors, only two have been studied extensively in meat products--ascorbic acid and α-tocopherol. Both are effective in reducing nitrosamine formation. Ascorbic acid is currently added to various cured meats, including bacon. Although α-tocopherol is currently not added to cured meats, a petition to permit such addition is being considered by the FDA (Hoffmann-La Roche, Inc., 1979). Because of their complementary solubility (ascorbate in aqueous, tocopherol in lipidic media), it is likely that a combination of ascorbate and α-tocopherol in nitrite-containing cured meats would provide effective protection against the exogenous formation of nitrosamines. Testing conducted shortly after processing has shown no impairment of the antibotulinal activity of nitrite by this combination. However, it is not clear whether such impairment would occur after increased storage, when residual nitrite has decreased.

The Committee recommends that further work be undertaken to ensure that the amount of ascorbate and α-tocopherol added to cured meats do not significantly impair the antibotulinal activity of nitrite, especially after increased storage times.

Because humans are exposed to *N*-nitroso compounds formed exogenously in the environment and endogenously in the body, the most effective means of reducing risks associated with exposure to these compounds would be to decrease exposures from both exogenous and endogenous sources. However, little is known about the factors influencing endogenous formation of *N*-nitroso compounds. Thus, the effect of incorporating nitrosation inhibitors into cured meats other than bacon on endogenous formation of *N*-nitroso compounds is unknown and such incorporation does not now appear to be warranted.

The Committee reiterates a recommendation made in its first report (National Academy of Sciences, 1981):

> The committee recommends that further research be conducted to study inhibition and catalysis of nitrosation reactions *in vivo*, specifically to determine the amount of nitrite that is destroyed in the human stomach and the extent to which nitrosation reactions are modified by the various inhibitors. Attention should also be directed toward interactions among inhibitors, catalysts, and other food-derived substances.

REFERENCES

Archer, M. C., S. R. Tannenbaum, T.-Y. Fan, and M. Weisman. 1975. Reaction of nitrite with ascorbate and its relation to nitrosamine formation. J. Natl. Cancer Inst. 54:1203-1205.

Bailey, M. E. 1980. Summary Report to NCR-95 Regional Committee, Colorado State University, December.

Bard, J., and W. E. Townsend. 1971. Meat curing. Pp. 452-483 in J. F. Price and B. S. Schweigert, eds. The Science of Meat and Meat Products. J. H. Freeman and Company, San Francisco.

Bharucha, K. R., C. K. Cross, and L. J. Rubin. 1976. Curing of meat using organic nitrites. U.S. Patent 3,966,974.

Bharucha, K. R., C. K. Cross, and L. J. Rubin. 1979. Mechanism of N-nitrosopyrrolidine formation in bacon. J. Agric. Food Chem. 27:63-69.

Bharucha, K. R., C. K. Cross, and L. J. Rubin. 1980. Long-chain acetals of ascorbic and erythorbic acid as antinitrosamine agents for bacon. Pp. 212-215 in Proceedings of the 26th European Meeting of Meat Research Workers, Volume 2, August 31-September 5, 1980, Colorado Springs, Colorado.

Brown, C. L., H. B. Hedrick, and M. E. Bailey. 1974. Characteristics of cured ham as influenced by levels of sodium nitrite and sodium ascorbate. J. Food Sci. 39:977-979.

Challis, B. C. 1973. Rapid nitrosation of phenols and its implications for health hazards from dietary nitrites. Nature 244:466.

Coleman, M. H. 1978. A model system for the formation of N-nitrosopyrrolidine in grilled or fried bacon. J. Food Technol. 13:55-69.

Coleman, M. H., R. S. Hannan, and D. R. D. Osborne. 1975. Cured meats. Can. Patent 964,107.

Dahn, H., L. Loewe, and C. A. Bunton. 1960. [In German; English Summary.] Uber die Oxydation von Ascorbinsäure durch salpetrige Säure. Teil VI: Ubersicht und Diskussion der Ergebnisse. Helv. Chim. Acta 43:320-333.

Davies, R., M. J. Dennis, R. C. Massey, and D. J. McWeeny. 1978. Some effects of phenol- and thiol-nitrosation reactions on *N*-nitrosamine formation. Pp. 183-197 in E. A. Walker, L. Griciute, M. Castegnaro, and R. E. Lyle, eds. Environmental Aspects of N-Nitroso Compounds. IARC Scientific Publication No. 19. International Agency for Research on Cancer, Lyon, France.

Douglass, M. L., B. L. Kabacoff, G. A. Anderson, and M. C. Cheng. 1978. The chemistry of nitrosamine formation, inhibition and destruction. J. Soc. Cosmet. Chem. 29:581-606.

Dudley, R. 1979. Quoted in, How to keep your bacon out of trouble. Meat Ind. 25(3):24-29.

Fiddler, W., J. W. Pensabene, E. G. Piotrowski, R. C. Doerr, and A. E. Wasserman. 1973. Use of sodium ascorbate or erythorbate to inhibit formation of *N*-nitrosodimethylamine in frankfurters. J. Food Sci. 38:1084.

Fiddler, W., J. W. Pensabene, E. G. Piotrowski, J. G. Phillips, J. Keating, W. J. Mergens, and H. L. Newmark. 1978. Inhibition of formation of volatile nitrosamines in fried bacon by the use of cure-solubilized α-tocopherol. J. Agric. Food Chem. 26:653-656.

Gough, T. A., K. S. Webb, and R. F. Coleman. 1978. Estimate of the volatile nitrosamine content of UK food. Nature 272:161-163.

Gray, J. I. 1976. *N*-Nitrosamines and their precursors in bacon: A review. J. Milk Food Technol. 39:686-692.

Gray, J. I., and L. R. Dugan, Jr. 1975. Inhibition of *N*-nitrosamine formation in model food systems. J. Food Sci. 40:981-984.

Gray, J. I., and C. J. Randall. 1979. The nitrite/*N*-nitrosamine problem in meats: An update. J. Food Prot. 42:168-179.

Greenberg, R. A. 1974. Ascorbate and nitrosamine formation in cured meats. Proceedings of the International Symposium on Nitrite in Meat Products, Zeist, the Netherlands.

Havery, D. C., T. Fazio, and J. W. Howard. 1978. Trends in levels of *N*-nitrosopyrrolidine in fried bacon. J. Assoc. Off. Anal. Chem. 61:1379-1382.

Herring, H. K. 1973. Effect of nitrite and other factors on the physico-chemical characteristics and nitrosamine formation in bacon. Pp. 47-60 in Proceedings of the Meat Industry Research

Conference, March 22-23, 1973. Meeting sponsored by the American Meat Science Association in cooperation with the American Meat Institute Foundation, both of Chicago, Illinois.

Hisatsune, I. C. 1961. Thermodynamic properties of some oxides of nitrogen. J. Phys. Chem. 65:2249-2253.

Hoffmann-La Roche, Inc. 1979. GRAS Affirmation Petition--dl-Alpha Tocopherol for Use in Bacon--GRASP 0G0265. Petition submitted December 20, 1979 to the Food and Drug Administration, Washington, D.C., filed Nov. 7, 1980. Fed. Reg. 45:74061.

Kamm, J. J., T. Dashman, A. H. Conney, and J. J. Burns. 1973. Protective effect of ascorbic acid on hepatotoxicity caused by sodium nitrite plus aminopyrine. Proc. Natl. Acad. Sci. USA 70:747-749.

Kamm, J. J., T. Dashman, A. H. Conney, and J. J. Burns. 1975. Effect of ascorbic acid on amine-nitrite toxicity. Ann. N. Y. Acad. Sci. 258:169-174.

Kamm, J. J., T. Dashman, H. Newmark, and W. J. Mergens. 1977. Inhibition of amine-nitrite hepatotoxicity by α-tocopherol. Toxicol. Appl. Pharmacol. 41:575-583.

Kelly, R. F. 1974. Processing meat products without nitrates or nitrites. Food Prod. Dev. 8:38-44.

Mergens, W. J., and H. L. Newmark. 1979. The use of alpha-tocopherol in bacon processing, an update. Pp. 79-84 in Proceedings of the Meat Industry Research Conference, March 29-30, 1979. American Meat Institute, Arlington, Virginia.

Mergens, W. J., and H. L. Newmark. 1980. Antioxidants as blocking agents against nitrosamine formation. Pp. 387-403 in M. Simic and M. Karel, eds. Autoxidation in Foods and Biological Systems. Plenum Press, New York and London.

Mergens, W. J., J. J. Kamm, H. L. Newmark, W. Fiddler, and J. Pensabene. 1978. Alpha-tocopherol: Uses in preventing nitrosamine formation. Pp. 199-212 in E. A. Walker, L. Griciute, M. Castegnaro, and R. E. Lyle, eds. Environmental Aspects of N-Nitroso Compounds. IARC Scientific Publication No. 19. International Agency for Research on Cancer, Lyon, France.

Mirvish, S. S. 1975. Formation of _N_-nitroso compounds: Chemistry, kinetics, and _in vivo_ occurrence. Toxicol. Appl. Pharmacol. 31: 325-351.

Mirvish, S. S. 1981. Inhibition of the formation of carcinogenic N-nitroso compounds by ascorbic acid and other compounds. Pp. 557-587 in J. H. Burchenal and H. F. Oettgen, eds. Cancer: Achievements, Challenges, and Prospects for the 1980s. Grune & Stratton, New York.

Mirvish, S. S., L. Wallcave, M. Eagen, and P. Shubik. 1972. Ascorbate-nitrite reaction: Possible means of blocking the formation of carcinogenic N-nitroso compounds. Science 177: 65-68.

Mottram, D. S. 1976. Increased N-nitrosamine formation in model fat systems containing sodium ascorbate. Pp. 135-136 in Proceedings of the Second International Symposium on Nitrite in Meat Products, September 7-10, 1976, Zeist, the Netherlands.

Mottram, D. S., and R. L. S. Patterson. 1977. The effect of ascorbate reductants on N-nitrosamine formation in a model system resembling bacon fat. J. Sci. Food Agric. 28:352-354.

Mottram, D. S., R. L. S. Patterson, D. N. Rhodes, and T. A. Gough. 1975. Influence of ascorbic acid and pH on the formation of N-nitrosodimethylamine in cured pork containing added dimethylamine. J. Sci. Food Agric. 26:47-53.

National Academy of Sciences. 1981. The Health Effects of Nitrate, Nitrite, and N-Nitroso Compounds. National Academy Press, Washington, D.C. 544 pp.

Newmark, H. L., and W. J. Mergens. 1981. α-Tocopherol (vitamin E) and its relationship to tumor induction and development. Pp. 127-168 in M. S. Zedeck and M. Lipkin, eds. Inhibition of Tumor Induction and Development. Plenum Press, New York and London.

Nitrite Safety Council. 1980. A survey of nitrosamines in sausages and dry-cured meat products. Food Technol. 34:45-51, 53, 103.

Pensabene, J. W., W. Fiddler, R. A. Gates, J. C. Fagan, and A. E. Wasserman. 1974. Effect of frying and other cooking conditions on nitrosopyrrolidine formation in bacon. J. Food Sci. 39:314-316.

Pensabene, J. W., J. I. Feinberg, E. G. Piotrowski, and W. Fiddler. 1979. Occurrence and determination of N-nitrosoproline and N-nitrosopyrrolidine in cured meat products. J. Food Sci. 44:1700-1702.

Pensabene, J. W., W. Fiddler, A. J. Miller, and J. G. Phillips. 1980. Effect of preprocessing procedures for green bellies on N-nitrosopyrrolidine formation in bacon. J. Agric. Food Chem. 28:966-970.

Pignatelli, B., M. Friesen, and E. A. Walker. 1980. The role of phenols in catalysis of nitrosamine formation. Pp. 95-109 in E. A. Walker, L. Griciute, M. Castegnaro, and M. Borzsonyi, eds. N-Nitroso Compounds: Analysis, Formation and Occurrence. IARC Scientific Publication No. 31. International Agency for Research on Cancer, Lyon, France.

Robach, M. C., J. L. Owens, M. W. Paquette, J. N. Sofos, and F. F. Busta. 1980. Effects of various concentrations of sodium nitrite and potassium sorbate on nitrosamine formation in commercially prepared bacon. J. Food Sci. 45:1280-1284.

Sebranek, J. G. 1979. Advances in the technology of nitrite use and consideration of alternatives. Food Technol. 33:58-62, 93.

Sen, N. P., and B. Donaldson. 1974. The effect of ascorbic acid and glutathione on the formation of nitrosopiperazines from piperazine adipate and nitrite. Pp. 103-106 in P. Bogovski and E. A. Walker, eds. *N*-Nitroso Compounds in the Environment. IARC Scientific Publication No. 9. International Agency for Research on Cancer, Lyon, France.

Sen, N. P., B. Donaldson, S. Seaman, J. R. Iyengar, and W. F. Miles. 1976. Inhibition of nitrosamine formation in fried bacon by propyl gallate and *L*-ascorbyl palmitate. J. Agric. Food Chem. 24:397-401.

Sen, N. P., B. Donaldson, S. Seaman, B. Collins, and J. Y. Iyengar. 1977. Recent nitrosamine analyses in cooked bacon. Can. Inst. Food Sci. Technol. J. 10:A13-A15.

Sen, N. P., S. Seaman, and W. F. Miles. 1979. Volatile nitrosamines in various cured meat products: Effect of cooking and recent trends. J. Agric. Food Chem. 27:1354-1357.

Tanaka, N. 1980. Effect of alpha-tocopherol on antibotulinal effectiveness of sodium nitrite in bacon. Attachment #37 to an Amendment to the GRAS affirmation petition--dl-Alpha-Tocopherol for Use in Bacon--GRASP 0G0265. Amendment submitted July 11, 1980 to the Food and Drug Administration, Washington, D.C.

Tanaka, K., K. C. Chung, H. Hayatsu, and T. Kada. 1978. Inhibition of nitrosamine formation *in vitro* by sorbic acid. Food Cosmet. Toxicol. 16:209-215.

U. S. Department of Agriculture. 1978. Nitrates, nitrites, and ascorbates (or isoascorbates) in bacon. Fed. Reg. 4:20992-20995.

Walters, C. L., M. W. Edwards, T. S. Elsey, and M. Martin. 1976. The effect of antioxidants on the production of volatile nitrosamines during the frying of bacon. Z. Lebensm. Unters. Forsch. 162:377-385.

CHAPTER 9

THE EFFECTS OF ALTERNATIVES TO NITRITE ON FLAVOR

Page

CHAPTER 9

THE EFFECTS OF ALTERNATIVES TO NITRITE ON FLAVOR

The question of replacement of nitrite in the generation of cured-meat flavor can be approached through both chemical studies and sensory evaluations. As compounds in cured meats are identified, especially those resulting from nitrite use, there is a tendency to smell or taste them, to see whether they might be the source of cured-meat flavor. This approach has some limitations.

Cured-meat flavor is most likely to be produced by volatile compounds and appreciated by the sense of smell. Although we typically speak of "tasting" flavors, smell is in reality the sense most involved in the identification of flavors (Mozell *et al.*, 1969). By taste, we perceive relatively few qualities--sweetness, saltiness, sourness, and bitterness are the only qualities universally accepted as related exclusively to taste. By the sense of smell, however, we perceive such a complex array of qualities that there is not even a universally accepted classification scheme (Cain, 1978). When food is placed in the mouth, volatiles rise in the rear of the oral cavity to the olfactory mucosa; when we eat, the sense of smell is effectively stimulated.

The sense of smell appears to process at least some aroma mixtures in a manner called "synthetic." A mixture of two aromas can produce an aroma that is perceived as qualitatively distinct from the aromas of the components (Amerine *et al.*, 1965, pp. 147-148). Thus, cured-meat "flavor" could be an aroma characteristic of a specific mixture of volatiles emitted by nitrite-cured meat, and a compound-by-compound search might miss such an aroma.

The Committee in its first report evaluated chemical aspects of cured-meat flavor formation and the contribution of nitrite to it. It concluded that cured-meat flavor is probably a composite sensation derived from the contributions of many odoriferous compounds. A positive contribution by nitrite to flavor cannot be specified in chemical terms, but for the reasons discussed above the inability to do this does not necessarily prove that none exists. Nitrite probably influences the flavor of cured meats by virtue of its antioxidant effects (Chapter 6). However, the Committee was unable to find any suggestions of compounds that had been promoted as alternatives to nitrite on the grounds that they produce the same profile of flavor-conferring chemicals as nitrite in cured meats.

Thus, this chapter focuses primarily on the evaluation of the flavor of products containing alternatives to nitrite that were identified as most promising in earlier chapters dealing with the other effects of nitrite (i.e., antimicrobial, antioxidant, and color-fixing). In addition, the contribution of nitrite to the flavor of various cured-meat products is also reviewed.

METHODS FOR COMPARING ALTERNATIVES WITH CONVENTIONAL NITRITE CURING

Sensory evaluation of the flavor contribution of alternatives to nitrite can be used to assess whether conventional nitrite curing and an alternative treatment yield the same sensory properties. Hedonic (preference) testing--another type of sensory evaluation--can be used to assess the palatability or acceptance of an alternative. The former criterion is the more stringent of the two.

Discriminative methods (which measure the ability to detect flavor differences) and descriptive methods (which measure particular sensory attributes) can both provide useful information. Discriminative methods include the classic triangle test commonly used in sensory evaluations of food products: a subject is given three samples, two of which are identical, and is asked to choose the odd one; above-chance choice behavior demonstrates that there is a sensory difference, but does not identify it. In descriptive methods, particular attributes (e.g., cured-meat flavor) are identified, and subjects are asked to classify samples on the basis of the magnitude of an identified attribute or asked to scale its magnitude in a variety of samples. The scaling procedure can use an ordinal approach (samples are ordered according to magnitude of the attribute), an interval approach (samples are scored according to magnitude based on prescribed scales), or a ratio approach (the magnitude of the attribute in each sample is measured).

Because the flavor contribution of nitrite depends on the nature of the meat product, comparisons using these methods must be made for each product if they are to be useful.

In discussing the flavor effects of nitrite and alternatives, some other issues are worthy of mention:

- Panel Selection. Panels used in the sensory evaluation of foods differ in their familiarity with test procedures, in their experience with foods, and possibly in sensory ability. Consumer panels are made up of consumers with no special training beyond the specific instructions required by the task. Trained panel members receive special training with food samples and may also be tested to ensure that their sensory abilities meet some standard. Training might involve tasting a variety of nitrite-cured meats, to gain familiarity with the attributes of cured-meat flavor. Such training often involves group discussions to develop descriptive terms that can be adopted by all participants. A person who consistently disagrees with the majority or who cannot perceive attributes generally agreed on is likely to be eliminated from such a panel.

Expert panels tend to be composed of persons with extensive experience in tasting the food products of interest (Amerine *et al.*, 1965, pp. 275-320). Obviously, the type of panel used will influence the results.

• Individual Differences and the Search for Cured-Meat Flavor. Identical food samples do not have the same flavors in the judgment of all persons (Amoore, 1977; Bartoshuk, 1979). This variation in sensory perception may contribute to the variability observed in sensory studies of cured-meat flavor. Chemosensory sensitivity must influence studies of the ability to discriminate nitrite-cured samples from those produced with alternatives. If insensitive persons are routinely eliminated from trained or expert panels, these panels could be expected to detect differences more reliably than would a random sample of the population. This is one reason why panel results do not always predict consumer behavior.

• Contributions to Cured-Meat Flavor by Factors Other Than Nitrite. In some products, cured-meat flavor can be produced by sodium chloride alone (Greene and Price, 1975; MacDougall *et al.*, 1975; Wasserman *et al.*, 1977; Williams and Greene, 1979). In some products, smoking of the meat also plays an important role in cured-meat flavor (Wasserman and Talley, 1972). The generation of smoke may result in the simultaneous generation of nitrogen oxides, which can influence color and flavor in the final product in a manner similar to that of added nitrite. The extent of such reactions is not known. Because of the many variables in the types of meat products and processing, and because addition of other flavoring agents, such as spices, may vary among product types, the contribution of nitrite to cured-meat flavor varies from one product to another. Thus, the contribution of alternative processing methods to flavor must be examined in each product.

THE ROLE OF NITRITE IN CURED-MEAT FLAVOR: A PRODUCT-BY-PRODUCT EVALUATION

Nitrite has been found to contribute to flavor in some products, and for others it has been shown that nitrite is not necessary for characteristic flavor. In cases other than those noted below, it is not possible to conclude with certainty whether nitrite makes a contribution. However, in ham-based products, a contribution similar to that in ham could be inferred.

Bacon

Bacon with an acceptable flavor can be prepared without nitrite. In studies with untrained panelists, Huhtanen *et al.* (1981) and Wasserman *et al.* (1977) found no difference between preference for nitrite-free and nitrite-cured bacon. These results were supported by Williams and Greene (1979), who found no difference between nitrite-free and nitrite-cured bacon in the amounts left on plates by breakfasting students.

Paquette *et al.* (1980) varied the amount of sodium nitrite in bacon samples from 0 to 120 mg/kg. Samples containing nitrite had a significantly more desirable flavor than did nitrite-free samples; however, there was no significant difference in desirability among samples with nitrite at different concentrations. Although the nitrite-free bacon had a less desirable flavor than the nitrite-cured bacon, it was still acceptable.

Both nitrite-free and nitrite-cured bacon samples in the studies cited above contained sodium chloride. Kimoto *et al.* (1976a,b) reported that sodium chloride is more important than nitrite to the flavor of bacon produced in the United States. The importance of sodium chloride was also demonstrated by MacDougall *et al.* (1975) in studies of English (Wiltshire) bacon. They compared the flavor of sodium chloride-free and nitrite-free bacon with that of bacon cured with various amounts of nitrite. Sodium chloride-free samples had almost no "bacon" flavor, but salted, nitrite-free bacon did have "bacon" flavor.

Frankfurters

During the preparation of frankfurters, salt and nitrite are added to meat emulsions with spices, sugars, and seasonings. Often, these products are also smoked.

Wasserman and Talley (1972) demonstrated that smoking is an important determinant of the flavor associated with frankfurters. Their panelists gave equivalent ratings of such flavor to nitrite-free and nitrite-cured samples when both were smoked. Unsmoked nitrite-cured frankfurters had more "frankfurter" flavor than unsmoked nitrite-free frankfurters.

Simon *et al.* (1973) found that all-beef frankfurters with no nitrite or with various concentrations of nitrite were judged to have equivalent flavor quality, whereas the flavor quality of half-pork, half-beef frankfurters varied directly with nitrite content.

The contribution of sodium chloride to the flavor of frankfurters has not been evaluated. However, Greene and Price (1975) found that salt was the major contributor to cured-meat flavor in samples of ground pork, whereas sodium nitrite alone (at 200 mg/kg) produced very little cured-meat flavor.

Ham

Brown *et al* (1974), MacDonald *et al.* (1980), and DuBose *et al.* (1981) confirmed the results of Greene and Price (1975) that sodium chloride can produce cured flavor. For example, MacDonald *et al.* showed that "nitrite-free" ham samples containing salt had a slight cured-ham flavor. This explains findings that acceptable country-style hams (i.e., dry-cured hams) can be produced without nitrite (Eakes and Blumer, 1975; Eakes _et al_., 1975; Kemp _et al_., 1974).

Nitrite does make a contribution to cured flavor in pickle-cured hams. MacDonald _et al_. (1980) cured hams with sodium nitrite at 50, 200, and 500 mg/kg in addition to sodium chloride. Their lowest nitrite addition was sufficient to produce a significant increase in cured-meat flavor, compared with that of samples containing only sodium chloride. It is likely that nitrite also contributes to the flavor of ham-based products.

THE EFFECTS OF ALTERNATIVE PROCESSING METHODS ON CURED-MEAT FLAVOR

In the products in which nitrite does not contribute significantly to flavor (e.g., bacon and some frankfurters), the major flavor assessment required will probably be to ensure that an antimicrobial alternative to nitrite does not introduce undesirable flavors. In products in which nitrite does contribute to flavor (e.g., ham), alternatives to nitrite may not produce the typical flavor. In this case, hedonic testing by representative consumers can be used to compare the acceptabilities of a conventional product and the same product prepared with an alternative.

Potassium Sorbate-Sodium Nitrite Combination in Bacon

A combination of potassium sorbate at 2,600 mg/kg and sodium nitrite at 40 mg/kg has been tested as an alternative to sodium nitrite at 120 mg/kg (currently in use) in bacon. Paquette _et al_. (1980) used a nine-point hedonic scale and consumer panels to evaluate the desirability of the color, appearance, texture, and flavor--as well as the overall desirability--of bacons prepared under normal commercial conditions and stored at $4^{o}C$ for various periods (7-63 d) after slicing and vacuum packaging. Bacons containing potassium sorbate at 2,600 mg/kg and sodium nitrite at 40 or 80 mg/kg were judged to be as desirable as bacon containing sodium nitrite at 120 mg/kg and no potassium sorbate. No adverse effects of testing were noted.

Berry and Blumer (1981) used an eight-point scale to evaluate the intensity of a variety of sensory attributes produced by smelling, tasting, and chewing bacon samples from four production plants. They found that the bacon prepared with potassium sorbate at 2,600 mg/kg plus sodium nitrite at 40 mg/kg produced an aroma during frying different from that of conventional bacon and a "chemical" flavor during tasting. In addition, some panelists reported a "prickly" mouth feeling from some of the sorbate-nitrite bacons. However, a later study (Robach and Adam, 1980) was unable to confirm that such reports were due to the presence of sorbate.

The results of Paquette *et al*. (1980) and Berry and Blumer (1981) provide an example of the difference between hedonic and sensory procedures. The data of Berry and Blumer (1981) show that panelists are able to perceive differences in flavor between conventionally cured bacon and bacon processed with a sorbate-nitrite combination, but the data of Paquette *et al*. (1980) lead to the conclusion that the two bacons are equally desirable. This highlights the need for both kinds of testing. The sensory data could be taken by themselves to imply that bacon processed with sorbate-nitrite is less desirable than conventionally cured bacon. The hedonic data could be taken by themselves to imply that the bacons are identical. Neither conclusion is warranted. The available data suggest that bacon processed with potassium sorbate at 2,600 mg/kg and sodium nitrite at 40 mg/kg does not have the same flavor as bacon cured with nitrite at 120 mg/kg; however, the flavors are equally desirable.

Lactic-Acid-Producing Organisms in Bacon and Country-Style Ham

The effects of the addition of lactic-acid-producing organisms on the sensory properties of bacon were examined by asking subjects to rate the magnitude of the difference between control bacon (prepared with sodium nitrite at 120 mg/kg) and experimental bacons (prepared with sodium nitrite at 40 or 120 mg/kg and inoculated with *Lactobacillus plantarum* or *Streptococcus faecalis*). Subjects were also asked to scale the intensity of the smokiness, saltiness, and off-flavor and to scale their overall preference for each bacon. The bacons were judged to be similar with regard to all measures except saltiness: the experimental bacons tended to be less salty than the control bacon (E. Traisman, Food Research Institute, University of Wisconsin, personal communication, 1981).

The effects of lactic-acid-producing organisms on the sensory properties of country-style hams were evaluated by asking trained panel members to judge a variety of attributes (e.g., saltiness, aged flavor, color, and bitterness) of ham slices from conventionally cured hams that had and had not been inoculated with *Pediococcus cerevisiae* (Bartholomew and Blumer, 1977). The hams were judged to be similar in flavor attributes.

Sodium Hypophosphite in Bacon

Data supplied to the Committee (J. J. Powers, University of Georgia, Atlanta, personal communication, 1981) suggest that bacon processed with sodium chloride and sodium hypophosphite at 3,000 mg/kg and bacon processed with sodium chloride, sodium hypophosphite at 1,000 mg/kg, and sodium nitrite at 40 mg/kg are judged to have flavor that is as desirable as that of conventionally processed bacon. Incidentally, bacon processed with sodium chloride alone was included as a control and was judged to have flavor as desirable as that of the other bacons. This corroborates the studies cited above that concluded that acceptable bacon can be prepared with sodium chloride alone.

Methylfumarate in Bacon

Data supplied by Huhtanen (U.S. Department of Agriculture Eastern Regional Research Laboratory, personal communication, 1981) on sensory evaluation suggest that bacon treated with methylfumarate at 1,250 mg/kg cannot be distinguished from control (conventional) bacon. In addition, hedonic scores for methylfumarate-treated bacon and conventional bacon were equivalent.

Irradiation and Reduced Nitrite in Ham, Corned Beef, and Bacon

In a series of studies, Wierbicki and colleagues systematically reduced the amounts of sodium nitrite added to a variety of irradiated meat products, to determine the minimal amount of sodium nitrite necessary to produce conventional color and flavor. Their hedonic evaluations used a nine-point scale to assess the acceptabilities of the odors and flavors of conventional products and those produced with reduced nitrite and irradiation. In addition, a nine-point hedonic scale was used to measure the overall preferences for the various products.

Ham with characteristic flavor and odor can be produced with reduced sodium nitrite (25 mg/kg) and irradiation. However, some nitrate must also be added to produce shelf-stable color (Wierbicki and Brynjolfsson, 1979; Wierbicki and Heiligman, 1974; Wierbicki *et al.*, 1977). The minimal sodium nitrate content is 25 mg/kg (Wierbicki and Brynjolfsson, 1979).

In irradiated corned beef, nitrite can be reduced to 50 mg/kg and still result in an acceptable product (Wierbicki and Brynjolfsson, 1979; Wierbicki *et al.*, 1977).

In irradiated bacon, nitrite can be reduced to 20 mg/kg and still result in an acceptable product. Irradiated bacon with no nitrite was also acceptable, but its overall preference score was significantly lower than that of bacon containing sodium nitrite at 20 mg/kg. The preference testing for specific attributes (color, odor, and flavor) showed that only color was significantly improved when nitrite was added (Wierbicki and Heiligman, 1980). This agrees with the above conclusion that nitrite is not necessary to produce typical "bacon" flavor.

It should be noted that the nitrite concentrations in these irradiated products were set on the basis of sensory characteristics. The nitrite content of irradiated meats could be reduced to zero if other methods could be used to produce desirable sensory properties.

CONCLUSIONS AND RECOMMENDATIONS

Whether nitrite contributes to the characteristic flavor of all cured products has not been determined. A contribution has been demonstrated for pickle-cured ham (and thus is probable in ham-based products). Acceptable dry-cured ham, bacon, and frankfurters (of some types) can be prepared without nitrite. Information is insufficient to conclude whether nitrite makes a contribution to the flavor of other products. Because smoking of meat may influence its flavor in a manner similar to nitrite, the contribution of smoking to the flavor of cured meat products should be investigated further.

Sensory evaluations of bacon containing the sorbate-nitrite combination and that containing lactic-acid-producing organisms and low nitrite have indicated that these products are acceptable. Sensory evaluation of irradiated bacon or corned beef containing low nitrite has indicated that these are also acceptable products. Irradiated low-nitrite ham is also acceptable, but requires some nitrate for shelf-stable color.

Other proposed alternatives have not been tested sufficiently for conclusions to be drawn. Additional sensory evaluation should accompany further testing of sodium hypophosphite and fumarate esters alone or in combination with low nitrite as possible antimicrobial alternatives to the current use of nitrite.

Undesired microbial proliferation in cured products can be inhibited by modifying product characteristics, such as a_w or pH (Chapter 5). To accompany research on the effect of such changes on antimicrobial protection, the Committee recommends that research be conducted on the effect of changes in the water activity or pH on sensory properties of various types of products. In the case of water activity, research should include modification of water activity by different means, e.g., the use of drying, salt, and other humectants.

REFERENCES

Amerine, M. A., R. M. Pangborn, and E. B. Roessler. 1965. Pp. 147-148 in Principles of Sensory Evaluation of Food. Academic Press, New York and London.

Amoore, J. E. 1977. Specific anosmia and the concept of primary odors. Chem. Senses Flavor 2:267-281.

Bartholomew, D. T., and T. N. Blumer. 1977. The use of a commercial Pediococcus cerevisiae starter culture in the production of country-style hams. J. Food Sci. 42:494-497.

Bartoshuk, L. M. 1979. Bitter taste of saccharin related to the genetic ability to taste the bitter substance 6-n-propylthiouracil. Science 205:934-935.

Berry B. W., and T. N. Blumer. 1981. Sensory, physical, and cooking characteristics of bacon processed with varying levels of sodium nitrite and potassium sorbate. J. Food. Sci. 46:321-327.

Brown, C. L., H. B. Hedrick, and M. E. Bailey. 1974. Characteristics of cured ham as influenced by levels of sodium nitrite and sodium ascorbate. J. Food Sci. 39:977-979.

Cain, W. S. 1978. History of research on smell. Pp. 197-229 in E. C. Carterette and M. P. Friedman, eds. Handbook of Perception, Volume 6A. Tasting and Smelling. Academic Press, New York, San Francisco, and London.

DuBose, C. N., A. V. Cardello, and O. Maller. 1981. Factors affecting the acceptability of low-nitrite smoked, cured ham. J. Food Sci. 46:461-463.

Eakes, B. D., and T. N., Blumer. 1975. Effect of various levels of potassium nitrate and sodium nitrite on color and flavor of cured lions and country-style hams. J. Food Sci. 40:977-980.

Eakes, B. D., T. N. Blumer, and R. J. Monroe. 1975. Effect of nitrate and nitrite on color and flavor of country-style hams. J. Food Sci. 40:973-976.

Greene, B. E., and L. G. Price. 1975. Oxidation-induced color and flavor changes in meat. Agric. Food Chem. 23:164-167.

Huhtanen, C. N., F. B. Talley, J. Feinberg, and J. G. Phillips. 1981. Flavor and antibotulinal evaluation of sorbic acid containing bacon. J. Food Sci. 1796-1800.

Kemp, J. D., J. D. Fox, and W. G. Moody. 1974. Cured ham properties as affected by nitrate and nitrite, and fresh pork quality. J. Food Sci. 39:972-976.

Kimoto, W. I., A. E. Wasserman, and F. B. Talley. 1976b. Sensory evaluation of flavor development in lean and adipose tissues of bacon. Lebensm. Wiss. Technol. 9:274-276.

MacDonald, B., J. I. Gray, D. W. Stanley, and W. R. Usborne. 1980. Role of nitrite in cured meat flavor: Sensory analysis. J. Food Sci. 45:885-888, 904.

MacDougall, D. B., D. S. Mottram, and D. N. Rhodes. 1975. Contribution of nitrite and nitrate to the colour and flavour of cured meats. J. Sci. Food Agric. 26:1743-1754.

Mozell, M. M., B. P. Smith, P. E. Smith, R. L. Sullivan, Jr., and P. Swender. 1969. Nasal chemoreception in flavor identification. Arch. Otolaryng. 90:131-137.

Paquette, M. W., M. C. Robach, J. N. Sofos, and F. F. Busta. 1980. Effects of various concentrations of sodium nitrite and potassium sorbate on color and sensory qualities of commercially prepared bacon. J. Food Sci. 45:1293-1296.

Robach, M. C., and K. Adam. 1980. Preliminary observations on human reactions related to sensory evaluation of cooked commercial bacon. J. Food Prot. 43:933-935.

Simon, S., D. E. Ellis, B. D. MacDonald, D. G. Miller, R. C. Waldman, and D. O. Westerberg. 1973. Influence of nitrite and nitrate curing ingredients on quality of packaged frankfurters. J. Food Sci. 38:919-923.

Wasserman, A. E., and F. Talley. 1972. The effect of sodium nitrite on the flavor of frankfurters. J. Food Sci. 37:536-538.

Wasserman, A. E., W. Kimoto, and J. G. Phillips. 1977. Consumer acceptance of nitrite-free bacon. J. Food Prot. 40:683-685.

Wierbicki, E., and A. Brynjolfsson. 1979. The use of irradiation to reduce or eliminate nitrite in cured meats (paper 8-2). Presented at the 25th European Meeting of Meat Research Workers, August 27-31, 1979, Budapest, Hungary.

Wierbicki, E., and F. Heiligman. 1974. Shelf stable cured ham with low nitrite-nitrate additions preserved by radappertization. Pp. 189-212 in B. Krol and B. J. Tinbergen, eds. Proceedings of the Second International Symposium on Nitrite in Meat Products, Zeist, PUDOC, Wageningen, the Netherlands.

Wierbicki, E., and F. Heiligman. 1980. Irradiated bacon without and with reduced addition of nitrite. Presented at the 26th European Meeting of Meat Research Workers, August 31-September 5, 1980, Colorado Springs, Colorado.

Wierbicki, E., F. Heiligman, and A. E. Wasserman. 1977. Irradiation as a conceivable way of reducing nitrites and nitrates in cured meats. Pp. 75-81 in B. J. Tinbergen and B. Krol, eds. Proceedings of the second International Symposium on Nitrite in Meat Products, Zeist, PUDOC, Wageningen, the Netherlands.

Williams, J. C., and B. E. Greene. 1979. Plate waste of bacon cured with and without sodium nitrite. J. Food Sci. 44:1260, 1262.

CHAPTER 10

THE TOXICOLOGY OF ALTERNATIVES TO NITRITE

Page

CHAPTER 10

THE TOXICOLOGY OF ALTERNATIVES TO NITRITE

The major adverse health effect arising from human exposure to nitrite in foods, such as cured meats, is the possible induction of cancer by N-nitroso compounds that may be formed in foods containing nitrate or nitrite or in the body after ingestion of such foods (National Academy of Sciences, 1981b). The preceding chapters evaluated possible alternatives to nitrite in cured meats that would result in the elimination or reduction of nitrite added to these products. In addition, because the most likely mechanism of the induction of cancer by nitrite is through the formation of nitrosamines, the feasibility of including specific inhibitors of nitrosation reactions in foods containing nitrite was discussed.

In this chapter, after the introduction of the Committee's criteria for the evaluation of toxicologic data, the data presented by the Committee in its first report (National Academy of Sciences, 1981b) on the adverse health effects of nitrite are summarized. The toxicity data on the agents or methods identified in the previous chapters as most promising substitutes for nitrite or as inhibitors of nitrosamine formation are then presented. These agents include ascorbic acid, lactic-acid-producing bacteria and lactic acid, glucono-∂-lactone, sodium chloride, wood smoke, sorbate, sodium hypophosphite, α-tocopherol, irradiation, and fumarate esters.

CRITERIA FOR EVALUATING TOXICOLOGY DATA

To determine whether the agents suggested as alternatives to nitrite are likely to cause adverse health effects in humans, a framework for assessing the available toxicology data is necessary. For the purposes of this discussion, the Committee has adopted a framework developed recently by the National Academy of Sciences (1981a). It was developed specifically for the assessment of the toxicity data on direct food additives based on exposures from intended use (in this case, ingestion, and not occupational or environmental exposure). The toxicity tests included in this framework are listed in Table 10-1.

Of course, the methods used in each of the tests listed--e.g., route of administration (i.e., gavage vs. administration in food or water), selection of species and strains, etc.--should be chosen on the basis of the nature of the substance, the anticipated human exposure, and the metabolism and pharmacokinetics of the substance in humans and various animals. Because duration of exposure can affect the degree of toxicity induced by a particular chemical, it is

TABLE 10-1

Tests and Data Required on Direct Food Additives[a]

Chemistry

Identification--e.g., molecular formula, purity, identification of contaminants

Physical data--e.g., melting and boiling points, water solubility, organic solubility, shelf-life

Toxicology

Acute oral toxicity--rodent

Subchronic oral toxicity--rodent: 14- or 28-d study

Subchronic oral toxicity--nonrodent: 90-d study

Subchronic oral toxicity--nonrodent: 6- to 12-mo study

Subchronic neurotoxicity: 90-d study[b]

Teratology--rodent, rabbit

Multigeneration reproduction study--rodent

Toxicokinetics[b]

Combined chronic toxicity-carcinogenicity--2 rodent species

Genetic toxicity

---

[a]Adapted from National Academy of Sciences, 1981a.
[b]Test not considered mandatory, but should be undertaken if results from other assays indicate a possible effect.

important to perform acute, subchronic, and chronic toxicity tests. For example, if a substance will be present consistently in foods and lifetime exposure of humans is likely, data from chronic feeding studies in animals should be used to assess the risk of chronic human exposure. Similarly, if it is likely that women of child-bearing age will be exposed to a substance, laboratory studies of reproductive injury are appropriate for assessing risks of fetotoxicity in humans. Several proposed alternatives to nitrite require combinations of chemicals, rather than a single chemical, and additional tests of the combinations are necessary to evaluate possible toxic interactions.

A complete evaluation of toxicology data must focus on the _amount_ of data available (i.e., whether all or most of the tests indicated have been performed) and on the _quality_ of the data (i.e., whether the assays have been performed according to accepted protocols). Because of the limited scope of this report, the Committee has not undertaken to answer the question concerning the quality of all reported data, especially for the substances classified as GRAS (generally recognized as safe). Instead, it suggests that an approach similar to that outlined by the National Academy of Sciences (1981a) could be used to answer the question. In this approach, standard protocols are selected by which the methods and results of each test can be measured (Table 10-2). In addition, it was suggested that, even if a test does not match a standard protocol, the data should be judged adequate if the tests meets "basic scientific criteria." The basic scientific criteria agreed to (National Academy of Sciences, 1981a) include the following:

- The purity and stability of the test substance and the dose, route, and duration of administration are clearly delineated.

- When possible, a species with metabolic pathways similar to those of humans is used.

- Controls are comparable with the test subjects in all respects except the treatment variable.

- End points answer the specific question addressed in the study and are sufficient to establish a dose-response relationship.

- An appropriate degree of statistical significance and adequate sample size are used.

In addition to data from laboratory studies, data on the extent of human exposure to a chemical, as well as data from epidemiologic studies, should be considered. In general, laboratory animal data are most valuable when the epidemiologic evidence is weak, nonspecific, relatively insensitive, or not obtainable. Conversely, good epidemiologic data minimize the need for animal data. Unfortunately, for most of the agents discussed in this chapter there are no reliable human epidemiologic data, and virtually all the toxicity information comes from studies conducted in animals and _in vitro_. In general, data from properly conducted animal studies are reasonably

TABLE 10-2

Reference Protocols for Toxicity Testing[a]

Test	Reference Sources
Acute oral toxicity--rodent	Interagency Regulatory Liaison Group, 1981
Subchronic oral toxicity--rodent: 14- or 28-d study	Organisation for Economic Co-operation and Development, 1981, pp. 407:1-8
Subchronic oral toxicity--nonrodent: 90-d study	Organisation for Economic Co-operation and Development, 1981, pp. 409:1-8
Subchronic oral toxicity--nonrodent: 6- to 12-mo study	Organisation for Economic Co-operation and Development, 1981, pp. 409:1-8
Subchronic neurotoxicity: 90-d study	Organisation for Economic Co-operation and Development, 1979, pp. 106-109
Teratology study--rodent, rabbit	U.S. Environmental Protection Agency, 1978
Multigeneration reproduction study--rodent	U.S. Environmental Protection Agency, 1978
Toxicokinetics	Organisation for Economic Co-operation and Development, 1981, pp. 415:1-5
Combined chronic toxicity-carcinogenicity--rodent	Organisation for Economic Co-operation and Development, 1981, pp. 453:1-15
Genetic toxicity	Organisation for Economic Co-operation and Development, 1979, pp. 114-116

[a]Adapted from National Academy of Sciences, 1981a.

predictive of the degree of risk to humans; however, because such laboratory investigations may be misleading with regard to target organ, potency, or type of toxic effect for individual substances, the judgment of scientific experts is an essential part of the analysis to ensure the proper use of all available data.

In the remainder of this chapter, the Committee reviews available toxicity data on the proposed alternatives and, where appropriate, draws conclusions on the adequacy of the data. In addition, recommendations concerning the need for further testing are made.

TOXICOLOGY OF NITRITE

Because nitrite is a prior-sanctioned substance for use in cured meats, its continued use is permitted, even though the data on its toxicity are incomplete and do not meet current requirements for approval of new food additives.

The following discussion summarizes the conclusions of the Committee given in its first report (National Academy of Sciences, 1981b).

Humans are exposed to nitrate and nitrite from a variety of sources. For the average U.S. resident, most nitrate comes from vegetables (87%), and most nitrite comes from cured meats (39%) and baked goods and cereals (34%). However, when the conversion of nitrate to nitrite in the human body is considered, most of the nitrite to which the average U.S. resident is exposed actually comes from vegetables (72%), and less than 10% comes from cured meats.

Although acute toxic effects, such as methemoglobinemia, can result from the ingestion of large amounts of nitrite or nitrate (especially in infants), the major possible adverse health effect that has been linked to nitrate and nitrite exposure is the induction of cancer; however, current evidence implicating nitrate, nitrite, and *N*-nitroso compounds in the development of cancer in humans is largely circumstantial. Epidemiologic studies have suggested a possible association between exposure to high concentrations of nitrate and nitrite and a high incidence of stomach and esophageal cancer in some parts of the world. However, in none of these studies was the exposure to nitrate, nitrite, or *N*-nitroso compounds actually measured in the persons who developed cancer. Moreover, other plausible causative agents were identified in many studies. Thus, it is not certain that nitrate or nitrite plays any role in the causation of these cancers. Studies conducted in animals to determine the carcinogenicity of nitrate and nitrite have also failed to provide sufficient evidence to permit a conclusion that these agents are directly carcinogenic. And, although nitrite is known to be mutagenic in microbial tests, its role in human cancer is most likely to be related to the formation of *N*-nitroso compounds.

Nitrate (after its reduction to nitrite *in vivo*) and nitrite can interact with amines and amides to produce nitrosamines and nitrosamides (*N*-nitroso compounds). These reactions can occur, under appropriate conditions, *in vitro* and *in vivo*. Thus, humans can be exposed to *N*-nitroso compounds that are formed in the environment or in the body after ingestion of nitrate, nitrite, and nitrosatable amino compounds.

Of the preformed nitrosamines present in the environment, most human exposure comes from nitrosamines in the air of some workplaces, tobacco smoke, cosmetics, pharmaceuticals, and pesticides. Of the dietary sources of preformed nitrosamines, the most important are cured meats, especially bacon, and beer. The formation of *N*-nitroso compounds *in vivo* has been well documented in laboratory animals. In humans, the evidence is sparse. However, one recent study showed that a noncarcinogenic nitrosamine was synthesized in a human subject after the ingestion of an amine (proline) and nitrate. In that experiment, the ingestion of a large excess of ascorbic acid or α-tocopherol effectively reduced the endogenous formation of the nitrosamine. Using data from this experiment, the Committee estimated in its first report that the amount of nitrosamines formed *in vivo* from the intake of nitrate and nitrite is roughly equivalent to the amount of preformed nitrosamines in the diet of the average person.

Many *N*-nitroso compounds have been shown to be mutagenic in microbial tests, either directly or with metabolic activation. Approximately 90% of the 300 *N*-nitroso compounds tested have been shown to be carcinogenic in one or more species of animals; this suggests that these compounds are likely to be carcinogenic in humans as well.

TOXICOLOGY OF ALTERNATIVES TO NITRITE

In this section, the toxicity data for alternatives judged by the Committee to be most promising in preceding chapters are reviewed. These alternatives are ascorbic acid and its salts, lactic-acid-producing bacteria and lactic acid, glucono-δ-lactone, sodium chloride, wood smoke, sorbic acid and potassium sorbate, sodium hypophosphite, α-tocopherol, irradiation, and fumarate esters.

For the alternatives that are classified as GRAS substances, the Committee has relied on the reviews done by the Select Committee on GRAS Substances of the Federation of American Societies for Experimental Biology (FASEB) for much of its toxicologic information. When more recent data were available, the Committee included them and discussed their implications. In addition, if information was available on alternatives to be used in combination with nitrite, the toxicology of the combination of agents is reviewed.

Alternatives Currently Permitted in Cured Meats

Of the alternatives listed above, several are currently permitted in the processing of cured-meat products--salts of ascorbic acid, lactic-acid-producing bacteria, glucono-δ-lactone, sodium chloride, and wood smoke.

Ascorbic Acid and Its Salts. L-Ascorbic acid, calcium and sodium L-ascorbate, erythorbic acid (D-isoascorbic acid), sodium erythorbate (sodium D-isoascorbate), and ascorbyl palmitate (palmitoyl L-ascorbate) are all GRAS substances. Ascorbate (in the form of sodium ascorbate or sodium erythorbate) is currently added to a variety of cured-meat products at concentrations of 200 mg/kg (in many products) to 550 mg/kg (in bacon). With the exception of bacon, ascorbate is added to cured-meat products to accelerate curing--especially color fixation (U.S. Department of Agriculture, 1970). The addition of ascorbate to bacon is required by federal regulation to inhibit the formation of nitrosamines. Because ascorbic acid may be used more consistently and at higher concentrations in products other than bacon in the future, the Committee has reviewed its toxicology to determine whether adverse health effects could result from the increased exposures.

Ascorbic acid (vitamin C) is an essential nutrient; all persons consume ascorbic acid as a normal constituent of foods. Many people also obtain it as a component of vitamin preparations or supplements in foods and beverages (Federation of American Societies for Experimental Biology, 1979a). The average daily intakes of ascorbic acid, erythorbic acid, and their salts have been estimated to total 47 mg (Federation of American Societies for Experimental Biology, 1979a). Although the concentration of sodium ascorbate or sodium erythorbate to be added to cured meats may vary from product to product, the maximal concentration is likely to be 550 mg/kg--the amount currently added to bacon. The Committee has estimated that, if this amount were added to all cured meats consumed in the United States, the per capita daily consumption would increase from 47 mg to 63.5 mg. This estimate is based on the assumption that the intake of cured meats is 30 g/d (National Academy of Sciences, 1981b, Chapter 5). This intake is less than the amount that would saturate the body pool in 95% of nonsmokers--100 mg/d (Kallner *et al.*, 1979).

Because sodium ascorbate is recommended, not as a replacement for nitrite, but as an inhibitor of nitrosation, it would be added to cured meats with the current amounts of sodium nitrite. In addition, α-tocopherol (at 500 mg/kg) would probably also be added to cured meat products, inasmuch as it is an active nitrosation inhibitor in lipids, whereas sodium ascorbate is active in aqueous media. The Committee is unaware of any toxicity tests of combinations of nitrite, sodium ascorbate, and α-tocopherol. Thus, the data that follow are derived from toxicity tests in which ascorbic acid or one of its salts was administered alone.

In mice, rats, and guinea pigs, the oral LD_{50} of sodium ascorbate was greater than 5.0 g/kg of body weight (Joint FAO/WHO Expert Committee on Food Additives, 1962). Subchronic oral-toxicity assays in the rat indicate that the maximal nontoxic dose is approximately 10 g/kg of body weight (Kieckebusch *et al.*, 1963).

Few studies of the toxic effects of ascorbic acid have been conducted over the entire lifetime in animals. In one long-term study with erythorbic acid administered in feed to rats for 2 yr at 1% of the diet, no pathologic lesions attributable to this exposure were detected (Lehman *et al.*, 1951).

Ascorbic acid administered for 10 consecutive days at doses of up to 550 mg/kg of body weight in pregnant rats and up to 520 mg/kg in pregnant mice was not teratogenic (Food and Drug Research Laboratories, Inc., 1975). However, tests for teratogenicity in chicks have yielded equivocal results (Hwang, 1974; Naber, undated; Verrett, 1977).

Ascorbic acid and erythorbic acid were not mutagenic in *in vitro* assays with *Salmonella typhimurium* and *Saccharomyces cerevisiae* with and without metabolic activation (Litton Bionetics, Inc., 1974a, 1975). Sodium erythorbate was not mutagenic in host-mediated assays in mice with *S. typhimurium* or *S. cerevisiae* D3 assays (SRI International, 1974). In contrast with these negative findings, Stich *et al.* (1976) reported that ascorbic acid in combination with copper salts did exhibit mutagenic activity in *S. typhimurium* assays. The biologic significance of this finding is unclear. More recent studies have also indicated that ascorbic acid may cause chromosomal aberrations (Stich *et al.*, 1980), sister chromatid exchanges (Galloway and Painter, 1979), and somatic mutations in cultured mammalian cells (Rosin *et al.*, 1980). In the latter study, high concentrations of ascorbate were required to induce mutations. Ascorbic acid has also been shown to inhibit the mutagenicity of a variety of known mutagens (Shamberger *et al.*, 1973), and Koropatnick and Stich (1980) recently demonstrated that ascorbate could increase or inhibit DNA damage induced by *N*-methyl-*N*'-nitro-*N*-nitrosoguanidine, depending on reaction conditions. In addition, epidemiologic evidence suggests that ascorbate may inhibit, rather than increase, carcinogenesis in humans (Bjelke, 1973; Mettlin *et al.*, 1981).

Lactic-Acid-Producing Bacteria and Lactic Acid. Lactic-acid-producing microorganisms, such as lactobacilli, may be added to bacon in amounts that are sufficient to reduce the pH--an effect that decreases nitrite concentration and thus inhibits nitrosamine formation (U.S. Department of Agriculture, 1979a). The cultures are also currently added as 0.5% of dry sausage, pork roll, Thüringer, Lebanon bologna, cervelet, and salami to establish the fermentative microflora rapidly and lower the pH. Lactic-acid-producing bacteria can lower the pH of cured-meat products, such as bacon, sufficiently to inhibit the outgrowth of *C. botulinum* spores (Chapter 5). Lactic-acid-producing bacteria, however, have not been approved as an additive to food products specifically for this purpose.

The types of bacteria added to meats are present naturally on the surface of meat and meat products and have been used as starter cultures in many other foods. Their safety has been established by long-term human use, and they are on the list of GRAS substances. If a new species of bacteria, other than those already added to meats or other foods, is to be used as an alternative to nitrite, a toxicologic evaluation would be required. However, much of the testing for antibotulinal activity discussed in this report has been done with species already used in meats or other foods (Smith and Polumbo, 1981).

Lactic acid is a GRAS substance and is currently added to a number of food items, including cheese, olives, frozen desserts, and wine (Federation of American Societies for Experimental Biology, 1978). Human exposure to lactic acid from food additives has been estimated to be 924 mg/person per day for persons 2 yr old and older (National Academy of Sciences, 1972). This estimate is based on production figures and should be taken as a _maximal_ daily intake. Such exposure is less than 1% of the endogenous exposure of humans to lactic acid produced metabolically by various body tissues--2 mg/kg per day, or approximately 120 mg/d for a 60-kg person (Federation of American Societies for Experimental Biology, 1978).

The addition of lactic acid to cured meats has been suggested (Chapter 5) as a means of reducing the pH to prevent spoilage and outgrowth of _C. botulinum_ spores. Although it has not been approved for this use in the United States, several European countries permit its addition to cured meats (Chapter 5). A recent review of the toxicology of lactic acid and calcium lactate by the Federation of American Societies for Experimental Biology (1978) revealed that, although acute and subchronic feeding studies have been conducted with lactic acid (Durlacher _et al._, 1946; Smyth _et al._, 1941), there have been no long-term feeding studies. Lactic acid is not mutagenic in _S. cerevisiae_ or _S. aureus_ (Litton Bionetics, Inc., 1976a,b).

Glucono-∂-lactone. Glucono-∂-lactone is a GRAS substance and is added to fermented sausages in this country at 5,000 mg/kg to reduce pH and retard microbial growth (U.S. Department of Agriculture, 1977). An acceptable daily intake for total gluconate (including glucono-∂-lactone) has been set at 50 mg/kg of body weight (Joint FAO/WHO Expert Committee on Food Additives, 1967). The average exposure in the United States to glucono-∂-lactone has been estimated to be 0.9 mg/d for a 60-kg adult (∿15 μg/kg of body weight), on the basis of annual use figures (National Academy of Sciences, 1972).

Acute toxicity studies conducted in cats and dogs (Chenoweth _et al._, 1941) revealed no adverse effects. In humans, high doses of glucono-∂-lactone (5-10 g/d) frequently caused stomach cramps or diarrhea (Chenoweth _et al._, 1941; Gold and Civin, 1939). No other adverse effects were detected. Glucono-∂-lactone was not mutagenic in tests with *Saccharomyces cereviseae* and *Salmonella typhimurium* (Litton Bionetics, Inc., 1974b).

The only long-term feeding study of glucono-∂-lactone reported was done by Van Logten *et al.* (1972). Rats were fed diets including 40% meat that contained 0.5% sodium nitrite and 1% glucono-∂-lactone, 0.2% sodium nitrite and 1% glucono-∂-lactone, or 1% glucono-∂-lactone. Rats fed 0.5% sodium nitrite and 1% glucono-∂-lactone had a lower growth rate than controls. No other adverse effects were seen in the test animals that were attributable to the various treatments. Administration of glucono-∂-lactone to pregnant mice, rats, hamsters, and rabbits did not cause any adverse effects (Food and Drug Research Laboratories, Inc., 1973c).

Sodium Chloride. Sodium chloride is used in the curing of meats to retard spoilage and impart flavor (Chapter 4). It has been suggested that increasing the concentration of sodium chloride in some products may reduce the need for sodium nitrite; however, the possible adverse health effects resulting from an increased sodium chloride intake are unknown. There is considerable evidence that such an increase may not be prudent (Federation of American Societies for Experimental Biology, 1979b).

Wood Smoke. Wood smoke has been used as a meat preservative for centuries. It has antimicrobial and antioxidant activities, as well as effects on color and flavor. The value of smoking as an alternative to nitrite is unclear, however, because nitrogen oxides in the smoke are most likely absorbed by the smoked meat and may act as nitrosating agents (Challis and Kyrtopoulos, 1978, 1979; Challis *et al.*, 1978) and produce *N*-nitroso compounds. In addition, nitrate and nitrite ions will be formed by hydrolysis of the oxides and deposited in the meat.

Alternatives Currently Approved for Use in Foods Other Than Cured Meats

Several proposed alternatives that are currently not added to cured meats are approved for addition to various other food products. These agents include sorbic acid and potassium sorbate, sodium hypophosphite, and α-tocopherol.

Sorbic Acid and Potassium Sorbate. Sorbic acid and potassium sorbate are treated collectively as "sorbate" in the discussion that follows.

Sorbate has been found to be a relatively safe food additive and is permitted in a variety of U.S. food products (Federation of American Societies for Experimental Biology, 1975a; Food and Drug Research Laboratories, Inc., 1973a; Furia, 1972; Sofos *et al.*,

1979). The Select Committee on GRAS Substances of the Federation of American Societies for Experimental Biology (1975a) estimated a per capita U.S. ingestion of about 25 mg/d. The Committee is unaware of more recent data that would allow an updated estimate of U.S. consumption. The Joint FAO/WHO Expert Committee on Food Additives (1974b) has estimated the acceptable daily intake of sorbic acid and sorbates (expressed as sorbic acid) to be 25 mg/kg of body weight per day--a very high acceptable daily intake among food preservatives. One estimate of sorbate intake (Federation of American Societies for Experimental Biology, 1975a) indicates that children 6-24 mo old may actually ingest this amount.

It has been proposed that sodium nitrite at 40 mg/kg (or potassium nitrite at 49 mg/kg) be used in bacon with potassium sorbate at 2,600 mg/kg (0.26%) and sodium ascorbate or sodium erythorbate at 550 mg/kg (U.S. Department of Agriculture, 1978a,b). This concentration of sorbate is high for a food additive, and up to 71% of added sorbate survives frying of bacon (Robach _et al._, 1980a); comparable concentrations of sorbate are added to some cheeses (Freese _et al._, 1973).

The average U.S. intake of cured meats is 30 g/d (Table 5-17 in National Academy of Sciences, 1981b), and bacon accounts for 19.5% of this (National Academy of Sciences, 1981b, Table 3-1), or 5.9 g of bacon per person per day. If one assumes that bacon contains potassium sorbate at 2,600 mg/kg (or 2,000 mg/kg expressed as sorbic acid) and that 71% of sorbate remains after frying, the average person would ingest approximately 8.4 mg of sorbate (expressed as sorbic acid) per day from bacon alone. Certainly, expansion of sorbate addition at this concentration to other cured-meat products would increase per capita consumption substantially, because an average person would consume cured meat at 30 g/d and thus would consume sorbate, expressed as sorbic acid, at 60 mg/d. High meat-eaters (National Academy of Sciences, 1981b, Table 10-9) would consume 240 mg/d of sorbate.

Because the Committee does not anticipate extensive expansion of sorbate addition to cured meats other than bacon and because the current proposal calls for potassium sorbate at 2,600 mg/kg plus sodium nitrite at 40 mg/kg and ascorbate at 550 mg/kg (U.S. Department of Agriculture, 1978a,b), the Committee has considered the possible health consequences of adding sorbate at this particular concentration. The toxicity data have been evaluated on sorbate alone and, where data are available, on the sorbate-nitrite mixture.

Toxicity of sorbate: The oral LD_{50} of sorbic acid in rats has been reported to be 7.36-10.50 g/kg of body weight (Deuel _et al._, 1954; Smyth and Carpenter, 1948). In rats, the LD_{50} of sodium sorbate (calculated as sorbic acid) was 5.9 g/kg (Deuel _et al._, 1954).

Sorbic acid has been fed to rats in a variety of studies. Doses ranged from 1.5 to 10% of the diet for 10 wk to 4 mo (Federation of American Societies for Experimental Biology, 1975a). The only effect seen was liver enlargement (Demaree *et al.*, 1955; Smyth and Carpenter, 1948). In another study, eight dogs were fed 1% and eight were fed 2% potassium sorbate in the diet for 3 mo (Deuel *et al.*, 1954). No adverse effects were detected.

No carcinogenic effect was demonstrated when diets containing 10% sorbic acid were fed to rats for 2 yr (Gaunt *et al.*, 1975). At this dose, the thyroid weight in males, the liver weight in both sexes, and the kidney, small intestine, and ovary weights in females were somewhat increased. Tests for teratogenicity in mice with potassium sorbate (460 mg/kg for 10 consecutive days) were negative (Food and Drug Research Laboratories, Inc., 1975).

Sorbates have been tested for mutagenicity in microbial and mammalian systems. Sorbate is inactive in a variety of bacterial systems (Difate, 1977; Hayatsu *et al.*, 1975; Kada, 1973; Khoudokormoff and Gist-Brocades, 1978a,b; Wood and Mergens, 1978); however, sorbic acid was active in the bacterial *rec* assay at 2-4 mg/ml (Namiki and Kada, 1975). In mammalian systems, potassium sorbate has been implicated in chromatid or chromosomal breaks, sister chromatid exchanges, and translocations in Chinese hamster cells *in vitro*; effective concentrations were high, but not unlike those which might be encountered after ingestion of sorbate in bacon (Abe and Sasaki, 1977; Ishidate and Odashima, 1977).

Sorbate, like many chemicals, infrequently causes allergic contact dermatitis. Fisher (1980) found that three patients with sorbic acid contact sensitivity who consumed food containing low concentrations of sorbate exhibited no flare of dermatitis on patch-test sites. No thorough studies seem to have been performed to assess human sensitization to sorbate at high concentrations, but orally administered sorbate is known to be rather rapidly metabolized in mammals (Federation of American Societies for Experimental Biology, 1975a; Food and Drug Research Laboratories, Inc., 1973a).

The Committee has found no reports of acute or subacute effects of ingested sorbic acid or sorbate in man, despite their wide use as food additives. Sorbate autooxidation can give rise to malonaldehyde and other carbonyls (Arya, 1980). The possible significance of malonaldehyde to human health is discussed in Chapter 6.

Sorbate-nitrite interactions: Although sorbate has passed numerous tests for safety when added to foods essentially free of nitrite, there is some concern regarding interactions between sorbate and nitrite when they are both added at relatively high concentrations to a single food item. Sorbate and nitrite are known to interact under a variety of conditions and to give rise to a number of chemical compounds, only some of which have been characterized.

A preliminary report of studies at the Food and Drug Administration indicated that commercially processed experimental bacon containing potassium sorbate at 2,600 mg/kg and nitrite at 40 mg/kg was as irritating for the hamster cheek pouch as was bacon processed with sodium nitrite alone at 120 mg/kg (Hinton et al., 1981). Both kinds of bacon were more irritating than was bacon free of either additive. The presence or absence of ascorbate in the curing mixtures was not reported. Similar results, indicating modest irritation by bacon containing conventional concentrations of nitrite and by nitrite-sorbate bacon in the hamster cheek pouch assay and on the skin of albino rabbits, have been noted elsewhere (Maibach, 1979a,b; Wolf, 1980; Wright, 1980a,b) and attributed mainly to the sodium chloride content of the samples tested. However, no similar studies in which sodium chloride was omitted from the processing mixture appear to have been performed.

The combination of sorbate and nitrite forms products with genotoxic activity for bacteria, particularly under conditions of acidic pH and a high molar ratio of nitrite to sorbate (Table 10-4). However, mutagens are still formed at a nitrite:sorbate ratio of 2:1 (Namiki et al., 1981) and 0.2:1 (Khoudokormoff and Gist-Brocades, 1978b).

Chemical species with DNA-damaging and mutagenic activity in bacterial test systems have been detected in sorbate-nitrite mixtures (Hayatsu et al., 1975; Kada, 1976; Khoudokormoff, 1978, 1981; Khoudokormoff and Gist-Brocades, 1978b; Kito and Namiki, 1978; Namiki and Kada, 1975; Namiki et al., 1980, 1981; Osawa et al., 1979, 1980). One product with genotoxic activity has been identified as ethylnitrolic acid (Namiki and Kada, 1975). Ethylnitrolic acid exhibits relatively low activity in the Salmonella/microsome assay (L. D. Kier, personal communication, 1981; Namiki et al., 1980), but is highly active in the rec assay (Kada, 1976; Namiki and Kada, 1975). Ethylnitrolic acid is 60 times as active as sorbic acid and 80 times as active as sodium nitrite in the bacterial rec assay (Namiki and Kada, 1975).

A second product of the reaction of sorbate and nitrite, compound Y, has been identified as 1,4-dinitro-2-methylpyrrole (Kito and Namiki, 1978; Osawa et al., 1979). Compound Y is active in the Salmonella/microsome assay (Namiki et al., 1980). The yield of mutants, about 14 per microgram with strain TA100, is in the range characteristic of other nitro compounds. A third product of the reaction of sorbate and nitrite, compound F, has been identified as a furoxan derivative of sorbate (Osawa et al., 1979). Compound F and its precursors do not appear to exhibit genotoxic activity when measured in the rec assay or in the Salmonella/microsome assay (Namiki et al., 1980, 1981). Another product of the reaction of sorbate and nitrite, compound B, has not been chemically characterized, but is also inactive in the Salmonella/microsome assay and only weakly active in the rec assay (Namiki et al., 1980, 1981).

TABLE 10-4

Mutagenicity of Nitrite-Sorbate Mixtures in Microbial Assays[a]

Reaction Conditions: Nitrite: Sorbate Molar Ratio	Reaction Conditions: pH	Result	Reference
10	2	+	Hayatsu *et al.*, 1975
1	?	+	Kada, 1973
0.2-0.8	3.0-6.5	+	Khoudokormoff and Gist-Brocades, 1978b
10	5.3	+	Wood and Mergens, 1978
4-16	1.5-5.0	+	Namiki *et al.*, 1979
5.6	?	+	Wasserman, 1979
0.025-2.5	5-7	-	Flowers *et al.*, 1979
0.025	1.3	-	Flowers *et al.*, 1979
0.17	?	-	Wasserman, 1979

[a]Adapted from Monsanto Company, 1980.

In contrast with the above findings, studies performed under different conditions, such as reduced molar concentrations or increased pH or both, have failed to show mutagenic activity for nitrite-sorbate mixtures (Table 10-4). However, conditions essential for bacterial inhibition by sorbate (optimal pH, 4.75) are also expected to provide an opportunity for mutagen formation (Khoudokormoff, 1981).

Apparently, detectable mutagenic activity in nitrite-sorbate mixtures is produced only under acidic conditions, optimally at a pH of 3.5-4.2, and is due mainly to the formation of compound Y and some ethylnitrolic acid (Namiki *et al.*, 1980, 1981). The latter compound predominates at lower pHs, but its formation requires high concentrations of reactants. Mutagens do not appear to be formed in typical bacon-curing brines or under simulated gastric conditions (Robach *et al.*, 1980b). However, these tests measured a combination of the formation and the persistence of mutagenic compounds under very specialized conditions. One important factor may have been the presence of ascorbate in the reaction mixtures. For example, Osawa *et al.* (1980) used preformed 1,4-dinitro-2-methylpyrrole and detected rapid degradation to a nonmutagenic derivative, 5-nitro-2-furaldehyde semicarbazone, in the presence of an eightfold molar excess of ascorbate near neutral pH, but no effect of ascorbate was noted at a pH of 1.5 or 3.5.

Based on the data reviewed above, it is likely that the known mutagenic reaction products do not form at detectable concentrations in raw bacon cured with sodium nitrite at 40 mg/kg plus potassium sorbate at 2,600 mg/kg plus sodium ascorbate or sodium erythorbate at 550 mg/kg. Because mutagenic activity is at least in part thermolabile (Namiki *et al.*, 1980), the persistence of these particular mutagenic products (even if preformed) in fried bacon is even less likely than their presence in raw bacon. However, mutagenic compounds could be formed during frying *in vivo* after ingestion. Although the sorbate reaction products are somewhat analogous to those expected from reaction of nitrite with other unsaturated aliphatic compounds (Mirna and Coretti, 1977; Walters *et al.*, 1979), which are naturally abundant in bacon (Benedict, 1980), these possible reactions may represent a potential health risk if the combination is used in a single product.

Sodium Hypophosphite. Several hypophosphites are GRAS substances; however, according to the Select Committee on GRAS Substances of FASEB (Federation of American Societies for Experimental Biology, 1977), no food manufacturer currently uses hypophosphites as additives.

The use of sodium hypophosphite in bacon at 1,000-3,000 mg/kg with sodium nitrite at 40 mg/kg has been suggested. Assuming an average intake of cured meats of 30 g/d for the U.S. population, the average intake of sodium hypophosphite as a food additive would increase from zero to 30-90 mg/d.

The acute oral toxicity of sodium hypophosphite has apparently not been measured (Federation of American Societies for Experimental Biology, 1977). In one short-term feeding study of calcium hypophosphite in rats, no adverse effects on growth were seen at various concentrations up to 4.3 g/kg of body weight over a 25-d test period (Meyer and Greenberg, 1959).

No long-term animal toxicity studies are available. There have also been no tests on hypophosphites for teratogenicity, reproductive effects, carcarcinogenicity, or mutagenicity.

α-Tocopherol. The tocopherols have vitamin E activity and are classified as GRAS substances. Tocopherols are present naturally in vegetable oils, cereals, nuts, and leafy vegetables. In addition, tocopherols (predominantly as α-tocopherol) are added to foods as antioxidants. The Select Committee on GRAS Substances of FASEB (Federation of American Societies for Experimental Biology, 1975b) concluded that, with the exception of infants receiving commercially prepared formulas, tocopherols added to foods contribute only slightly to total tocopherol intake. The FASEB committee estimated that tocopherols added to foods are ingested at an average of approximately 0.6 mg/d. The tocopherol intake from the diet has been estimated at 7-9 mg/d (Bieri and Evarts, 1973; Bunnell *et al*., 1965).

The Accepted Daily Intake (ADI) for tocopherols (expressed as α-tocopherol for a 60-kg person) developed by the Joint FAO/WHO Expert Committee on Food Additives (1974a) is 120 mg/d. Assuming that α-tocopherol is added to cured-meat products at 500 mg/kg and that 30 g of these products is consumed daily, the Committee estimates that an additional intake of 15 mg/d would result. This is roughly twice the current estimated total dietary intake of tocopherol, but still considerably less than the ADI estimated by the Expert Committee on Food Additives.

This value for daily intake is misleading, however, in the case of bacon. Only 7.5% of the α-tocopherol added to bacon is present in the edible portion after frying (Mergens and Newmark, 1979). Thus, the concentration of α-tocopherol in fried bacon is ∿41 mg/kg. If the average intake of bacon is 5.9 g/d, the intake of tocopherol from this source would be only 0.24 mg/d.

The LD_{50} values for acute oral toxicity of the tocopherols are relatively high--over 2,000 mg/kg in rabbits, and over 4,000 mg/kg in rats and mice (Federation of American Societies for Experimental Biology, 1977).

Excessive intake of tocopherols leads to hypervitaminosis E in several species. Conditions described as hypervitaminosis E have been reported in guinea pigs (Prosperi and Borselli, 1955), hamsters (Czyba, 1966), and chicks (March *et al*., 1973). However, there is no convincing evidence that high intake of tocopherol leads to hypervitaminosis E in humans; similarly, physiologic benefits from the addition of large amounts of vitamin E to the diet have not been demonstrated (Federation of American Societies for Experimental Biology, 1977).

The administration of α-tocopherol to pregnant mice, rats, hamsters, and rabbits had no teratogenic effects (Food and Drug Research Laboratories, Inc., 1973b). α-Tocopherol was toxic when administered in the air cell of chick embryos at doses of 48 mg/kg and higher; however, no teratogenic effects were seen at doses of 0-120 mg/kg (Reid, 1975).

No long-term toxicity studies that examined the carcinogenicity of α-tocopherol have been reported. However, numerous studies have assessed the effect of α-tocopherol on the carcinogenicity of other agents. α-Tocopherol was reported to retard the growth of spontaneously occurring mammary tumors in mice (Dobrovolskäia-Zaradskaïa and Adamova, 1945), reduce the mortality from sarcomas induced by benzo-[a]pyrene in mice (Bonmassar *et al*., 1968), and reduce the incidence of tumors caused by 3-methylcholanthrene in mice (Shamberger, 1970). However, other studies have revealed no apparent effect of α-tocopherol in inhibiting carcinogenicity (Demole, 1939; Wattenberg, 1972). α -Tocopherol was not mutagenic in bacterial mutagenicity tests (SRI International, 1979) and has been reported to reduce carcinogen-induced chromosomal damage (Shamberger *et al*., 1973).

Irradiation. Although irradiation technology is well developed to the point of producing aesthetically acceptable foods that are microbiologically safe and nonradioactive, the only approved uses for irradiation of food are for sprout inhibition of potatoes and insect disinfestation of wheat and flour (Food and Drug Administration, 1981). Also, although they are approved, there has been no commercial application of these uses of irradiation.

The Food and Drug Administration is developing proposals for the regulation of irradiated foods. As a first step in this process, the Food and Drug Administration (1981) published the report of an internal agency task force that makes the following recommendations:

- That food irradiated at doses of 100 kilorads (krads) or less be considered wholesome and safe for human consumption.

- That food irradiated at doses exceeding 100 krads be subject to toxicity testing consisting of a battery of four short-term mutagenicity tests and two 90-d feeding studies (one in rodents, one in nonrodent mammals).

- That a food class making up no more than 0.01% of the daily diet and irradiated at 5 Mrads or less be considered safe for human consumption without toxicity testing.

The major concern regarding safety has been the lack of adequate information on the chemical products of irradiation. Convincing evidence regarding the wholesomeness of many irradiated foods has been obtained (Anonymous, 1981; Barna, 1979; Reber *et al*., 1966). Toxicity studies have been conducted on irradiated chicken and fish (as well as a wide variety of other foodstuffs). Long-term studies in two rodent species, a three-generation reproduction study in rodents, and a 1-yr study in dogs on the effects of ingesting irradiated chicken showed no adverse effects (Joint FAO/IAEA/WHO Expert Committee on the Wholesomeness of Irradiated Food, 1977). Short-term studies in mice, rats, and dogs fed either mixed eviscerated cod (a nonfatty fish) and redfish (a fatty fish) or other fish showed no adverse effects (Joint FAO/IAEA/WHO Expert Committee on the

Wholesomeness of Irradiated Food, 1977). The mixed cod and redfish material has also been tested in rats in long-term, multigeneration reproduction, dominant-lethal, and cytogenetic studies. No adverse long-term effects were seen.

The Joint FAO/IAEA/WHO Expert Committee on the Wholesomeness of Irradiated Food (1977) has stressed that toxicologic approaches to the assessment of irradiated food must be based on the concept of food irradiation as a process, and not as an additive. That is, the Joint Committee concluded that the toxicologic evaluation of irradiated food differs from the safety evaluation of chemicals added to food. In its report, the Joint Committee stated: "It is impracticable to exaggerate the feeding levels of irradiated foods in animal studies beyond a modest degree, nor is it appropriate to exaggerate the radiation dosage much beyond that to be used in practice. Either of these practices gives rise to effects which are not relevant to the toxicological potential of the irradiated food. The evaluation of the wholesomeness of irradiated foods therefore poses problems of a different kind from those encountered with food additives or contaminants and it consequently requires a different approach."

Although the Joint Committee has also recently concluded that "the irradiation of any food commodity up to an overall average dose of 10 kGy [1 Mrad] presents no toxicological hazard; hence, toxicological testing of foods so treated is no longer required" (Anonymous, 1981), and the doses used to treat cured meats are within this range (Chapter 5), the present Committee is unaware of any published reports on long-term toxicity tests of irradiated foods containing nitrite, i.e., no long-term testing has been done on irradiated cured products.

Alternatives Currently Not Approved for Use in Foods

Fumarate Esters. Several chemical analogues of methyl and ethyl fumarate esters are being studied for antibotulinal activity (Huhtanen, U.S. Department of Agriculture, personal communication, 1981). In limited tests conducted with monomethylfumarate and monoethylfumarate at 1,250 mg/kg in bacon, promising results were achieved. However, information on the metabolism and toxicity of these compounds is not available.

CONCLUSIONS AND RECOMMENDATIONS

On the basis of the evidence presented, the Committee has concluded that the addition of ascorbate to cured meats at 550 mg/kg is unlikely to result in adverse health effects in humans. However, additional tests of the mutagenicity of ascorbate, especially in combination with α-tocopherol and sodium nitrite, may be warranted. If mutagenic products are formed, animal toxicity assays of this combination of agents may also be prudent.

If the species of lactic-acid-producing bacteria to be used as an alternative to nitrite differs from those already in use in foods, the Committee recommends that the necessary toxicologic evaluation be conducted. The health implications of adding lactic acid to meats in sufficient amounts to retard outgrowth of *C. botulinum* spores are unclear, because information on chronic toxicity is absent and because natural *in vivo* production of lactic acid far exceeds the amount that would be ingested in the meat products.

The Committee recommends that increased concentrations of sodium chloride not be considered as an alternative to nitrite at this time, because of possible adverse effects of the resulting increase in sodium intake. Similarly, because smoking may foster the formation of nitrosamines in meats, the Committee recommends that it not be considered as an alternative to nitrite.

Although sorbate alone does not appear to be toxic, the combination of sorbate and nitrite results, under some conditions, in the formation of mutagenic reaction products. The Committee recommends that the following assays be conducted with characterized sorbate-nitrite reaction products, to permit a more complete determination of the possible adverse effects of the combination of nitrite and sorbate in meats: short-term assays with mammalian (including human) cell-culture systems, acute and chronic toxicity tests, and toxicokinetics studies.

Owing to the paucity of toxicity data on sodium hypophosphite, the Committee recommends that the necessary toxicity tests be undertaken to ascertain its safety for use as a substitute for nitrite in cured meats.

Although the addition of α-tocopherol to cured meats at a concentration of 500 mg/kg is unlikely to have adverse effects on human health, mutagenicity tests and acute and subchronic toxicity tests of combinations of α-tocopherol, ascorbate, and nitrite should be conducted to determine whether toxic reaction products are formed.

The Committee recommends that toxicity studies of irradiation focus on the effects of irradiation in combination with the addition of low concentrations of nitrite (40 mg/kg).

If other substances, such as the fumarate esters, are found to be effective antimicrobial agents in a variety of meat products when tested under commercial conditions, the Committee recommends that toxicity tests be conducted on these agents as well.

REFERENCES

Abe, S., and M. Sasaki. 1977. Chromosome aberrations and sister chromatid exchanges in Chinese hamster cells exposed to various chemicals. J. Natl. Cancer Inst. 58:1635-1641.

Anonymous. 1981. Wholesomeness of Irradiated Food. Summary of the Report of a Joint FAO/IAEA/WHO Expert Committee, Geneva, October 27-November 3, 1980. Food Irradiat. Inf. 11:8-20.

Arya, S. S. 1980. Stability of sorbic acid in aqueous solutions. J. Agric. Food Chem. 28:1246-1249.

Barna, J. 1979. Compilation of bioassay data on the wholesomeness of irradiated food items. Acta Aliment. 8:205-315.

Benedict, R. C. 1980. Biochemical basis for nitrite inhibition of Clostridium botulinum in cured meat. J. Food Prot. 43:877-891.

Bieri, J. G., and R. P. Evarts. 1973. Tocopherols and fatty acids in American diets. The recommended allowance for vitamin E. J. Am. Diet. Assoc. 62:147-151.

Bjelke, E. 1973. Epidemiologic Studies of Cancer of the Stomach, Colon, and Rectum; With Special Emphasis on the Role of Diet, Volumes 1-4. Ph.D. Thesis from the University of Minnesota, Minneapolis, Minnesota. Abstract available in Diss. Abstr. Int. B: 34(8):3879B-3880B.

Bonmassar, E., R. Dallavalle, and G. Giuliani. 1968. Influsso della vitamina E e del gallato di propile sulla cancerogenesi da benzopirene nel topo. Arch. Ital. Pathol. Clin. Tumori 11:245-250.

Bunnell, R. H., J. Keating, A. Quaresimo, and G. K. Parman. 1965. Alpha-tocopherol content of foods. Am. J. Clin. Nutr. 17: 1-10.

Challis, B. C., and S. A. Kyrtopoulos. 1978. The chemistry of nitroso compounds. Part 12. The mechanism of nitrosation and nitration of aqueous piperidine by gaseous dinitrogen tetraoxide and dinitrogen trioxide in aqueous alkaline solutions. Evidence of molecular isomers of dinitrogen tetraoxide and dinitrogen trioxide. J. Chem. Soc. Perkin Trans. II: 1296-1302.

Challis, B. C., and S. A. Kyrtopoulos. 1979. The chemistry of nitroso-compounds. Part II. Nitrosation of amines by the two phase interaction of amines in solution with gaseous oxides of nitrogen. J. Chem. Soc. Perkin Trans. I:299-304.

Challis, B. C., A. Edwards, R. R. Hunma, S. A. Kyrtopoulos, and J. R. Outram. 1978. Rapid formation of N-nitrosamines from nitrogen oxides under neutral and alkaline conditions. Pp. 127-142 in E. A. Walker, L. Griciute, M. Castegnaro, and R. E. Lyle, eds. Environmental Aspects of N-Nitroso Compounds. IARC Scientific Publication No. 19. International Agency for Research on Cancer, Lyon, France.

Chenoweth, M. B., H. Civin, C. Salzman, M. Cohn, and H. Gold. 1941. Further studies on the behavior of gluconic acid and ammonium gluconate in animals and man. J. Lab. Clin. Med. 26:1574-1582.

Czyba, J. C. 1966. Effets de l'hypervitaminose E sur le testicule du hamster dore. C. R. Soc. Biol. 160:765-768.

Demaree, G. E., D. W. Sjogren, B. W. McCashland, and F. P. Cosgrove. 1955. Preliminary studies on the effect of feeding sorbic acid upon the growth, reproduction, and cellular metabolism of albino rats. J. Am. Pharm. Assoc. Sci. Ed. 44:619-621.

Demole, V. 1939. Zur Frage der tumorbildenden Wirkung des Weizenkeimöls. Z. Vitaminforsch. 8:341-347.

Deuel, H. J., Jr., R. Alfin-Slater, C. S. Weil, and H. F. Smyth, Jr. 1954. Sorbic acid as a fungistatic agent for foods. I. Harmlessness of sorbic acid as a dietary component. Food Res. 19:1-12.

Difate, V. G. 1977. Microbiological mutagenicity evaluation of sorbic acid/sodium nitrite reaction products. A report of August 31, 1977, to Dr.Robert Angelotti, Chairman, Expert Panel on Nitrite and Nitrosamines, U. S. Department of Agriculture, Washington, D.C.

Dobrovolskaïa-Zavadskaïa, N., and N. Adamova. 1945. L'Influence de L' α-tocopherol administré per as sur la croissance des tumeurs chez les Souris.

Durlacher, S. H., W. Harrison, Jr., and D. C. Darrow. 1946. The effects of calcium chloride and of calcium lactate administered by gavage. Yale J. Biol. Med. 18:135-143.

Federation of American Societies for Experimental Biology. 1975a. Evaluation of the Health Aspects of Sorbic Acid and Its Salts as Food Ingredients. A report prepared for the Bureau of Foods, Food and Drug Administration, Washington, D.C., under Contract No. FDA 223-75-2004. Life Sciences Research Office, Federation of American Societies for Experimental Biology, Bethesda, Maryland. Available from National Technical Information Service as PB-262 663. 17 pp.

Federation of American Societies for Experimental Biology. 1975b. Evaluation of the Health Aspects of Tocopherols and Alpha-Tocopheryl Acetate as Food Ingredients. A report prepared for the Bureau of Foods, Food and Drug Administration, Washington, D.C., under Contract No. FDA 223-75-2004. Life Sciences Research Office, Federation of American Societies for Experimental Biology, Bethesda, Maryland. Available from National Technical Information Service as PB-262 653. 30 pp.

Federation of American Societies for Experimental Biology. 1977. Evaluation of the Health Aspects of Hypophosphites as Food Ingredients. A report prepared for the Bureau of Foods, Food and Drug Administration, Washington, D. C., under Contract No. FDA 223-75-2004. Life Sciences Research Office, Federation of American Societies for Experimental Biology, Bethesda, Maryland. Available from National Technical Information Service as PB-274 476. 10 pp.

Federation of American Societies for Experimental Biology. 1978. Evaluation of the Health Aspects of Lactic Acid and Calcium Lactate as Food Ingredients. A report prepared for the Bureau of Foods, Food and Drug Administration, Washington, D.C., under Contract No. FDA 223-75-2004. Life Sciences Research Office, Federation of American Societies for Experimental Biology, Bethesda, Maryland. 21 pp.

Federation of American Societies for Experimental Biology. 1979a. Evaluation of the Health Aspects of Ascorbic Acid, Sodium Ascorbate, Calcium Ascorbate, Erythorbic Acid, Sodium Erythorbate, and Ascorbyl Palmitate as Food Ingredients. A report prepared for the Bureau of Foods, Food and Drug Administration, Washington, D.C., under Contract No. FDA 223-75-2004. Life Sciences Research Office, Federation of American Societies for Experimental Biology, Bethesda, Maryland. 48 pp.

Federation of American Societies for Experimental Biology. 1979b. Evaluation of the Health Aspects of Sodium Chloride and Potassium Chloride as Food Ingredients. A report prepared for the Bureau of Foods, Food and Drug Administration, Washington, D.C., under Contract No. FDA 223-75-2004. Life Sciences Research Office, Federation of American Societies for Experimental Biology, Bethesda, Maryland. Available from National Technical Information Service as PB-298 139/AS. 69 pp.

Fisher, A. A. 1980. Cutaneous reactions to sorbic acid and potassium sorbate. Cutis 25:350, 352, 423.

Flowers, L. J., J. A. Grindley, L. D. Kier, F. A. Kulik, L. H. Lee, J. Sonstein, and W. D. Ross. 1979. Final report on *Salmonella* Mutagenicity Assays of Sorbate/Nitrite Reaction Mixtures

(EHL Study #790037). Report submitted to the Food and Drug Administration, Washington, D.C., by Monsanto Research Corporation, Dayton, Ohio.

Food and Drug Administration. 1981. Policy for irradiated foods: Advance notice of proposed procedures for the regulation of irradiated foods for human consumption. Fed. Reg. 46:18992-18994.

Food and Drug Research Laboratories, Inc. 1973a. Scientific Literature Reviews on Generally Recognized As Safe (GRAS) Food Ingredients--Sorbic Acid and Derivatives. A report prepared for the Food and Drug Administration, Washington, D.C. Food and Drug Research Laboratories, Inc. 92 pp.

Food and Drug Research Laboratories, Inc. 1973b. Teratologic Evaluation of FDA 71-58 (*dl*-alpha-Tocopheryl Acetate) in Mice, Rats, Hamsters, and Rabbits. A report prepared under DHEW contract no. FDA 71-260. Maspeth, New York. 44 pp.

Food and Drug Research Laboratories, Inc. 1973c. Teratologic Evaluation of FDA 71-72 (Glucono-delta-lactone). A report prepared for the Food and Drug Administration under Contract No. FDA 71-260. Waverly, New York. 56 pp.

Food and Drug Research Laboratories, Inc. 1975. Teratologic Evaluation of FDA 71-65 (Ascorbic Acid) in Mice and Rats. A report prepared under DHEW contract No. FDA 223-74-2176. Waverly, New York. 26 pp. plus appendixes.

Freese, E., C. W. Sheu, and E. Galliers. 1973. Function of lipophilic acids as antimicrobial food additives. Nature (London) 241:321-325.

Furia, T. E., ed. 1972. Sorbic acid and its salts. Pp. 129-137 in CRC Handbook of Food Aditives, 2nd Ed. CRC Press, Cleveland, Ohio.

Galloway, S. M., and R. B. Painter. 1979. Vitamin C is positive in the DNA synthesis inhibition and sister-chromatid exchange tests. Mutat. Res. 60:321-327.

Gaunt, I. F., K. R. Butterworth, J. Hardy, and S. D. Gangolli. 1975. Long-term toxicity of sorbic acid in the rat. Food Cosmet. Toxicol. 13:31-45.

Gold, H., and H. Civin. 1939. Gluconic acid as a urinary acidifying agent in man. J. Lab. Clin. Med. 24:1139-1146.

Hayatsu, H., K. C. Chung, T. Kada, and T. Nakajima. 1975. Generation of mutagenic compound(s) by a reaction between sorbic acid and nitrite. Mutat. Res. 30:417-419.

Hinton, D. M., E. A. Brouwer, F. J. Vocci, A. Joshi, G. Yang, and D. I. Ruggles. 1981. Pilot studies of the membrane toxicity and chemical reactivity of combinations of sorbate and nitrite in a processed food. Fed. Proc. 40:878. Abstract 3682.

Hwang, U. K. 1974. Investigation of the toxic and teratogenic effects of GRAS substances to the developing chicken embryo: Erythorbic acid. A report to M. J. Verrett, Food and Drug Administration, Washington, D.C. St. Louis University School of Medicine, St. Louis, Missouri. 8 pp.

Interagency Regulatory Liaison Group. 1981. Recommended Guideline for Acute Oral Toxicity Testing in Rodents. Interagency Regulatory Liaison Group, Testing Standards and Guidelines Work Group, Washington, D.C. 8 pp.

Ishidate, M., Jr., and S. Odashima. 1977. Chromosome tests with 134 compounds on Chinese hamster cells _in vitro_--a screening for chemical carcinogens. Mutat. Res. 48:337-354.

Joint FAO/IAEA/WHO Expert Committee on the Wholesomeness of Irradiated Food. 1977. Wholesomeness of Irradiated Food. FAO Food and Nutrition Series No. 6. WHO Technical Report Series No. 604. Food and Agriculture Organization of the United Nations, Rome, Italy. 44 pp.

Joint FAO/WHO Expert Committee on Food Additives. 1962. Evaluation of the Toxicity of a Number of Antimicrobials and Antioxidants. 6th Report. Food and Agriculture Organization of the United Nations, Rome, Italy.

Joint FAO/WHO Expert Committee on Food Additives. 1967. Toxicological Evaluation of Some Antimicrobials, Antioxidants, Emulsifiers, Stabilizers, Flour-Treatment Agents, Acids and Bases. FAO Nutr. Meet. Rep. Ser. (40 A, B, C), pp. 139-140. Food and Agriculture Organization of the United Nations, World Health Organization, Geneva, Switzerland.

Joint FAO/WHO Expert Committee on Food Additives. 1974a. α-Tocopherol and mixed tocopherols concentrate. Pp. 208-209 in Toxicological Evaluation of Some Food Additives Including Anticaking Agents, Antimicrobials, Antioxidants, Emulsifiers and Thickening Agents. World Health Organization Food Additives Series No. 5. Food and Agriculture Organization of the United Nations, Rome, Italy; and World Health Organization, Geneva, Switzerland.

Joint FAO/WHO Expert Committee on Food Additives. 1974b. Sorbic acid and its calcium, potassium, and sodium salts. Pp. 121-129 in Toxicological Evaluation of Some Food Additives Including Anticaking Agents, Antimicrobials, Antioxidants, Emulsifiers and Thickening Agents. World Health Organization Food Additives Series No. 5. Food and Agriculture Organization of the United Nations, Rome, Italy; and World Health Organization, Geneva, Switzerland.

Kada, T. 1973. DNA-damaging products from reaction between sodium nitrite and sorbic acid. Annu. Rep. Natl. Inst. Genet. 24: 43-44.

Kada, T. 1976. Mutagenicity and carcinogenicity screening of food additives by the rec-assay and reversion procedures. Pp. 105-115 in R. Montesano, H. Bartsch, L. Tomatis, and W. Davis, eds. Screening Tests in Chemical Carcinogenesis. IARC Scientific Publication No. 12. International Agency for Research on Cancer, Lyon, France.

Kallner, A., D. Hartmann, and D. Hornig. 1979. Steady-state turnover and body pool of ascorbic acid in man. Am. J. Clin. Nutr. 32:530-539.

Khoudokormoff, B. 1981. Evaluation of the mutagenicity of sorbic acid-sodium nitrite reaction products. Food Cosmet. Toxicol. 19:405-406.

Khoudokormoff, B., and N. V. Gist-Brocades. 1978a. A modification of Kada's rec-assay technique for detection of low level mutagenicity, and its application to some food additives. Paper presented at the International Symposium on Chemical Toxicology of Food, June 8-10, 1978, Milan, Italy.

Khoudokormoff, B., and N. V. Gist-Brocades. 1978b. Potential carcinogenicity of some food preservatives in the presence of traces of nitrite. Mutat. Res. 53:129-130.

Kieckebusch, W., W. Griem, and K. Lang. 1963. Untersuchungen über die chronische Toxizität der Ascorbinsäure bei der Ratte. Z. Ernaehr. 4:5-14.

Kito, Y., and M. Namiki. 1978. A new N-nitropyrrole,1,4-dinitro-2-methylpyrrole, formed by the reaction of sorbic acid with sodium nitrite. Tetrahedron 34:505-508.

Koropatnick, D. J., and H. F. Stich. 1980. The modifying effect of sodium ascorbate on DNA damage and repair after N-methyl-N'-nitro-N-nitrosoguanidine treatment *in vivo*. Biochem. Biophys. Res. Commun. 92:292-298.

Lehman, A. J., O. G. Fitzhugh, A. A. Nelson, and G. Woodard. 1951. The pharmacological evaluation of antioxidants. Adv. Food Res. 3:197-208.

Litton Bionetics, Inc. 1974a. Mutagenic Evaluation of Compound FDA 71-66, Erythorbic Acid. Report prepared for the Food and Drug Administration, Washington, D.C., under DHEW Contract No. FDA 223-74-2104. Kensington, Maryland. 34 pp.

Litton Bionetics, Inc. 1974b. Mutagenic Evaluation of Compound FDA 71-72 Glucono-Delta-Lactone. Report prepared for the Food and Drug Administration, Washington, D.C.,under Contract No. 223-74-2104. Kensington, Maryland. 37 pp. Available from National Technical Information Service as PB 245-438.

Litton Bionetics, Inc. 1975. Mutagenic Evaluation of Compound FDA 71-65, Ascorbic Acid. Report prepared for the Food and Drug Administration, Washington, D.C., under DHEW Contract No. FDA 223-74-2104. Kensington, Maryland. 41 pp.

Litton Bionetics, Inc. 1976a. Mutagenic Evaluation of Compound FDA 75-19, Lactic Acid, USP. Report prepared for the Food and Drug Administration, Washington, D.C., under DHEW Contract No. 223-74-2104. Kensington, Maryland. 20 pp. with appendixes.

Litton Bionetics, Inc. 1976b. Mutagenic Evaluation of Compound FDA 75-20, Calcium Lactate, USP. Report prepared for the Food and Drug Administration, Washington, D.C., under DHEW Contract No. 223-74-2104. Kensington, Maryland. 20 pp. with appendixes.

Maibach, H. 1979a. Primary Skin Irritation Test. Bacon Study I. In Monsanto Company Petition to the FDA--Sorbate Use in Meat Products and Poultry, November 1980.

Maibach, H. 1979b. Primary Skin Irritation Test. Bacon Study II. In Monsanto Company Petition to the FDA--Sorbate Use in Meat Products and Poultry, November 1980.

March, B. E., E. Wong, L. Seier, J. Sim, and J. Biely. 1973. Hypervitaminosis E in the chick. J. Nutr. 103:371-377.

Mergens, W. J., and H. L. Newmark. 1979. The Fate of dl-Alpha-Tocopherol in Bacon upon Frying. Attachment 25 in the GRAS Affirmation Petition. dl-Alpha-Tocopherol in Bacon. Submitted December 20, 1979, to the Food and Drug Administration, Washington, D.C., by Hoffmann-LaRoche, Inc., New Jersey.

Mettlin, C., S. Graham, R. Priore, J. Marshall, and M. Swanson. 1981. Diet and cancer of the esophagus. Nutr. Cancer 2:143-147.

Meyer, A. E., and J. Greenberg. 1949. Value of calcium hypophosphite and other calcium compounds as calcium supplements in calcium-low diets. Proc. Soc. Exp. Biol. Med. 71:40-43.

Mirna, A., and K. Coretti. 1977. Inhibitory effect of nitrite reaction products and of degradation products of food additives. Pp. 39-45 in B. J. Tinbergen and B. Kroll, eds. Proceedings of the 2nd International Symposium on Nitrite in Meat Products, September 7-10, 1976, Zeist, PUDOC, Wageningen, the Netherlands.

Monsanto Company. 1980. GRAS Affirmation Petition. Sorbic acid and Potassium Sorbate for Use as a Preservative in Meat Products, Fresh Poultry, and Poultry Products. Document submitted by the Monsanto Company to the Food and Drug Administration, Washington, D.C.

Naber, E. C. Undated. Investigations on the toxic and teratogenic effects of GRAS substances on the developing chick embryo. A report prepared under FDA Contract No. FDA 72-343. Ohio State University, Columbus, Ohio. 94 pp.

Namiki, M., and T. Kada. 1975. Formation of ethylnitrolic acid by the reaction of sorbic acid with sodium nitrite. Agric. Biol. Chem. 39:1335-1336.

Namiki, M., S. Udaka, T. Osawa, K. Tsuji, K. Namiki, and T. Kada. 1979. Mutagen formation by sorbic acid and nitrite reaction conditions and products. Paper presented at the ACS/CSJ Chemical Congress, April 1-6, 1979, Honolulu, Hawaii.

Namiki, M., S. Udaka, T. Osawa, K. Tsuji, and T. Kada. 1980. Formation of mutagens by sorbic acid-nitrite reaction: Effects of reaction conditions on biological activities. Mutat. Res. 73:21-28.

Namiki, M., T. Osawa, H. Ishibashi, K. Namiki, and K. Tsuji. 1981. Chemical aspects of mutagen formation by sorbic acid-sodium nitrite reaction. J. Agric. Food Chem. 29:407-411.

National Academy of Sciences. 1972. A Comprehensive Survey of Industry on the Use of Food Chemicals Generally Recognized As Safe (GRAS). A report prepared for the Food and Drug Administration by the Committee on Food Protection, Division of Biology and Agriculture, National Research Council. National Academy of Sciences, Washington, D.C. 41 pp.

National Academy of Sciences. 1981a. Strategies to Determine Needs and Priorities for Toxicity Testing. Volume 1. Design. A report by the Steering Committee on Identification of Toxic and Potentially Toxic Chemicals for Consideration by the National Toxicology Program. National Academy of Sciences, Washington, D.C. 143 pp.

National Academy of Sciences. 1981b. The Health Effects of Nitrate, Nitrite, and N-Nitroso Compounds. National Academy Press, Washington, D.C. 544 pp.

Organisation for Economic Co-operation and Development. 1979. OECD Short-Term and Long-Term Toxicology Groups. Final Report. Organisation for Economic Co-operation and Development, Paris, France. 185 pp.

Organisation for Economic Co-operation and Development. 1981. OECD Guidelines for Testing of Chemicals. Organisation for Economic Co-operation and Development, Paris, France. 700 pp.

Osawa, T., Y. Kito, M. Namiki, and K. Tsuji. 1979. A new furoxan derivative and its precursors formed by the reaction of sorbic acid with sodium nitrite. Tetrahedron Lett. 45:4399-4402.

Osawa, T., H. Ishibashi, M. Namiki, and T. Kada. 1980. Desmutagenic actions of ascorbic acid and cysteine on a new pyrrole mutagen formed by the reaction between food additives; sorbic acid and sodium nitrite. Biochem. Biophys. Res. Commun. 95:835-841.

Prosperi, P., and L. Borselli. 1955. Studioistologico sulla ipervitaminosi E nella cavia. Riv. Clin. Pediatr. 56:378-392.

Reber, E. F., K. Raheja, and D. Davis. 1966. Wholesomeness of irradiated foods--an annotated bibliography. Fed. Proc. 25:1529-1579.

Reid, B. L. 1975. Vitamin E (dl-α-tocopheryl acetate): Toxicity and teratogenicity studies in avian embryos. Report prepared under FDA Contract No. 72-345. Department of Poultry Science, University of Arizona, Tucson, Arizona. 18 pp.

Robach, M. C., J. L. Owens, M. W. Paquette, J. N. Sofos, and F. F. Busta. 1980a. Effects of various concentrations of sodium nitrite and potassium sorbate on nitrosamine formation in commercially prepared bacon. J. Food Sci. 45:1280-1284.

Robach, M. C., V. G. DiFate, K. Adam, and L. D. Kier. 1980b. Evaluation of the mutagenicity of sorbic acid-sodium nitrite reaction products produced in bacon-curing brines. Food Cosmet. Toxicol. 18:237-240.

Rosin, M. P., R. H. C. San, and H. F. Stich. 1980. Mutagenic activity of ascorbate in mammalian cell cultures. Cancer Lett. 8:299-305.

Shamberger, R. J. 1970. Relationship of selenium to cancer. I. Inhibitory effect of selenium on carcinogenesis. J. Natl. Cancer Inst. 44:931-936.

Shamberger, R. J., F. F. Baughman, S. L. Kalchert, C. E. Willis, and G. C. Hoffman. 1973. Carcinogen-induced chromosomal breakage decreased by antioxidants. Proc. Natl. Acad. Sci. 70:1461-1463.

Smith, J. L., and S. A. Polumbo. 1981. Microorganisms as food additives. J. Food Prot. 44:936-955.

Smyth, H. F., Jr., and C. P. Carpenter. 1948. Further experience with the range finding test in the industrial toxicology laboratory. J. Ind. Hyg. Toxicol. 30:63-68.

Smyth, H. F., Jr., J. Seaton, and L. Fischer. 1941. The single dose toxicity of some glycols and derivatives. J. Ind. Hyg. Toxicol. 23:259-268.

Sofos, J. N., F. F. Busta, and C. E. Allen. 1979. Botulism control by nitrite and sorbate in cured meats: A review. J. Food Prot. 42:739-770.

SRI International. 1974. Study of mutagenic effects of sodium erythorbate (FDA no. 71-68). Compound Report No. 4. Report prepared for the Food and Drug Administration, Washington, D.C., under Contract No. FDA 73-215. Stanford Research Institute, Menlo Park, California. 55 pp. with appendixes.

SRI International. 1979. Microbial mutagenesis testing of substances. Compound Report F76-057, dl-alpha-tocopherol, General Mills Chemicals, Lot No. 767-1181, CAS #000059029. Report prepared for the Food and Drug Administration, Washington, D.C., under Contract No. 323-77-2118. Menlo Park, California. 13 pp.

Stich, H. F., J. Karim, J. Koropatnick, and L. Lo. 1976. Mutagenic action of ascorbic acid. Nature 260:722-724.

Stich, H. F., L. Wei, and R. F. Whiting. 1980. Chromosome aberrations in mammalian cells exposed to vitamin C and multiple vitamin pills. Food Cosmet. Toxicol. 18:797-801.

U.S. Department of Agriculture. 1970. P. 15590 in Meat Inspection Regulations. Revision pursuant to Wholesome Meat Act. Fed. Reg. 35:15552-15617.

U.S. Department of Agriculture. 1977. Task Force Report: The Staphylococcal Enterotoxin Problem in Fermented Sausage. Food Safety and Quality Service, U.S. Department of Agriculture, Washington, D.C. 13 pp. plus appendixes.

U.S. Department of Agriculture. 1978a. Federal meat inspection regulations. Substances used in preparation of bacon. Proposed rulemaking. Fed. Reg. 43:21007.

U.S. Department of Agriculture. 1978b. Substances used in preparation of bacon. Revision of proposed rule. Fed. Reg. 43: 39119-39120.

U.S. Department of Agriculture. 1979a. Acid-producing microorganisms in meat products for nitrite dissipation: Final rule. Fed. Reg. 44(31):9372.

U.S. Department of Agriculture. 1979b. Substances used in preparation of bacon. Fed. Reg. 44:28331.

U.S. Environmental Protection Agency. 1978. Pesticide programs: Proposed guidelines for registering pesticides in the U.S.; Hazard evaluation: Humans and domestic animals. Fed. Reg. 43(163):37336-37403.

Van Logten, M. J., E. M. Den Tonkeliar, R. Kroes, J. M. Berkvens, and G. J. van Esch. 1972. Long-term experiment with canned meat treated with sodium nitrite and glucono-∂-lactone in rats. Food Cosmet. Toxicol. 10:475-488.

Verrett, M. J. 1977. Investigations of the Toxic and Teratogenic Effects of GRAS Substances to the Developing Chicken Embryo: Calcium Ascorbate. Food and Drug Administration, Washington, D.C. 6 pp.

Walters, C. L., R. J. Hart, and S. Perse. 1979. The possible role of lipid pseudonitrosites in nitrosamine formation in fried bacon. Z. Lebensm. Unters. Forsch. 168:177-180.

Wasserman, A. E. 1979. Mutagenicity Tests of Nitrite/Sorbate Brine Solutions and Bacon. Report submitted to R. Engel, U.S. Department of Agriculture, Washington, D.C.

Wattenberg, L. W. 1972. Inhibition of carcinogenic and toxic effects of polycyclic hydrocarbons by phenolic antioxidants and ethoxyquin. J. Natl. Cancer Inst. 48:1425-1430.

Wolf, J. E. 1980. The Effects of Bacon on the Hamster Cheek Pouch. In Monsanto Company Petition to the FDA--Sorbate Use in Meat Products and Poultry, November 1980.

Wood, A. W., and W. J. Mergens. 1978. Evaluation of the Mutagenic Potential of Mixtures of Sorbic Acid and Sodium Nitrite. Document submitted by Hoffmann-La Roche, Inc., to the U.S. Department of Agriculture in response to Federal Register Notice--Substances Used in Preparation of Bacon-- May 16, 1978.

Wright, P. L. 1980a. Hamster Cheek Pouch Tests with Bacon Samples Containing Sodium Nitrite and Potassium Sorbate. In Monsanto Company Petition to the FDA--Sorbate use in Meat Products and Poultry, November 1980.

Wright, P. L. 1980b. Summary of Primary Skin Irritation Tests with Bacon Samples Containing Nitrite, Potassium Sorbate, or Combinations of the Two as Preservatives. In Monsanto Company Petition to the FDA--Sorbate Use in Meat Products and Poultry, November 1980.

CHAPTER 11

DEVELOPMENT AND EVALUATION OF ALTERNATIVES TO NITRITE

CHAPTER 11

DEVELOPMENT AND EVALUATION OF ALTERNATIVES TO NITRITE

This chapter presents the results of the Committee's deliberation on two aspects of its charge: the long-term approaches to the development of alternatives that the Committee feels are most likely to be profitable and the most appropriate ways in which to evaluate the efficacy of alternatives. Throughout this chapter, the emphasis is on antimicrobial alternatives, because from the point of view of health protection the need for such alternatives is most obvious.

DEVELOPMENT OF ALTERNATIVES TO NITRITE: LONG-TERM APPROACHES

Antimicrobial Alternatives

On the basis of its assessment of the antimicrobial effects of nitrite (National Academy of Sciences, 1981, Chapter 3), the Committee concluded that research intended to develop replacements should focus primarily on agents or techniques for inhibiting the growth of spore-forming microorganisms, because of their role in the causation of botulism and product spoilage.

Much effort has been devoted in the last decade to identifying potential antibotulinal or anticlostridial replacements for nitrite. This effort has met with little success--perhaps not surprisingly, in view of the formidable restrictions inherent in finding a nitrite replacement. Such an agent not only must be efficacious and have negligible resistance liability, but must be stable during processing and storage, and, because of the widespread anticipated exposure of the human population to the agent, it must not pose any toxicologic threat, either directly or through its metabolites or reaction products.

A number of the qualities expected of an antimicrobial food additive are similar to those required of therapeutic antimicrobials. The Committee therefore examined approaches to the development of specific antimicrobial agents that are now being used, to a limited extent, in the pharmaceutical industry. These approaches, outlined in more detail in Appendix B, capitalize on biochemical differences between humans and the microorganisms to be inhibited. A "target"--i.e., a component, such as an enzyme, involved in a process essential to microbial survival or proliferation--is chosen that is absent in the human or is of a different structure (see also Appendix C), and compounds that interact specifically with the target, thereby inhibiting growth, are developed or selected.

Application of this approach was considered by the Committee to be rational in the search for antimicrobial alternatives to nitrite. In examining its feasibility, the Committee focused on *Clostridium botulinum* as the pathogenic microorganism of major concern in cured-meat products and as a suitable representative of the clostridia and spore-forming bacteria in general.

Some distinct processes or events can be recognized in the progression that culminates in a food that is spoiled or contaminated with botulinum toxin. These are shown in Table 11-1, with examples of factors that influence them. The Committee believes that efforts to arrest the progression toward toxin synthesis or spoilage should focus on the early events in the sequence (before cell multiplication), because these events have no metabolic equivalents in humans. Additionally, after outgrowth has occurred, increasingly large numbers of cells would have to be inhibited to arrest the progression, and, if the inhibition fails, the duration of temperature abuse required for toxin synthesis would be correspondingly shorter.

In an effort to identify targets in *C. botulinum* to which the approach outlined in Appendix B could be applied, the Committee examined the literature on *C. botulinum* spores and the physiologic processes by which they develop into replicating vegetative cells. However, the literature is meager. As pointed out by Smoot and Pierson (in press), work on the spores of aerobic microorganisms (such as bacilli), because of its greater convenience, has taken precedence over that on the anaerobic clostridia. Moreover, within the latter group, *C. botulinum* has been neglected in favor of studies on nonpathogenic species. The Committee therefore concluded that too little is now known about the development of *C. botulinum* spores into cells to justify focusing on one possible target in a program to design an antibotulinal agent. Nonetheless, the Committee remains convinced that the approach is a rational one for developing alternatives to nitrite.

To gather the information that would facilitate identification of suitable targets, such as the critical steps in the spore-to-cell transition, the Committee recommends concurrent pursuit of several types of investigation.

Basic research into the physiology of *C. botulinum* (and other microorganisms not desired in meat products) is needed to identify peculiarities in their metabolic processes that might profitably be exploited to control their growth. Preliminary indications of some possibilities are noted in Appendix C. In the case of *C. botulinum* and other spore-formers, studies should focus primarily on spore outgrowth, but the injury of spores (Foegeding and Busta, 1981) or their dormancy, resistance to attack, activation, and germination should also be investigated. The control of vegetative-cell multiplication of both spore-forming and non-spore-forming microorganisms also needs some further investigation.

One means of investigating processes involved in the spore-to-cell transition is examination of the mechanism of action of agents that promote particular steps in the sequence, e.g., the effect of

TABLE 11-1

Stages in the Production of Toxin from *C. botulinum* Spores[a]

Process	Adversely Affected By	Promoted By	Suitability As Primary Target[b]
Maintenance of spore	Heat (with low pH), carbonyl compounds, radiation	Dehydration, optimum pH optimal pH	+++++
Activation of spores	Refrigeration	Mild heating, acids, thiol compounds	++++
Germination of spores	Sorbate, D-alanine, NaCl, pH extremes, D-cysteine, extreme temperature	L-Alanine, L-cysteine, carbon dioxide or bicarbonate, calcium lactate	+++[c]
Outgrowth of spores	NO_2, NaCl, low a_w, low pH, low temperature	--	+++++
Cellular multiplication	Low or high temperature, low a_w, low pH, radiation	--	++
Toxin synthesis	--	Some carbohydrates, cations	+
Maintenance of toxin activity	Heat above 80°C, proteolysis	--	-[d]

[a]Sources: Ando, 1973, 1974; Benedict, 1980; Sakaguchi, 1979; Smith, 1977; Smoot and Pierson, in press. The list of agents promoting or adversely affecting processes is not intended to be exhaustive.

[b]+++++ = high, + = low, - = unsuitable. See text.

[c]A number of alternative "routes" of germination probably exist (Foegeding and Busta, 1981), rendering blocking of all necessary to inhibit the process.

[d]Toxin activity may be destroyed by some adverse influences, e.g., cooking, but products may be cooked insufficiently to destroy toxin or not at all.

carbon dioxide or L-alanine on spore germination. However, more basic studies are also needed--e.g., on the metabolic pathways involved in these processes and on the enzymes. Other studies are needed to determine the mechanisms whereby some microorganisms (or their spores) are more resistant to particular stresses than are other species. For example, why are Bacillus stearothermophilus spores considerably more resistant to heat than C. botulinum spores?

Investigation of the modes of action of the known inhibitors of C. botulinum is desirable, because it would probably reveal potential target sites for a program to develop agents that inhibit spore-forming microorganisms. It is of little consequence that a known agent may be a relatively weak inhibitor--it is still likely to be a useful tool that could help in identifying a point of susceptibility on which further studies could capitalize. Some possibilities along these lines are discussed in Appendix C. Further investigations of the mechanisms of action of nitrite against C. botulinum and other organisms should be given high priority, because they would yield information on one target site that is already known to be highly susceptible to attack. The reasons for the resistance of some microorganisms to nitrite also need investigation.

In studies of both physiologic processes and mechanisms of inhibition, the Committee believes that experiments should be designed to yield information that is related closely to the microenvironments of the meat, poultry, and fish products to which nitrite is now added.

In recent decades, there has been little funding for research on spore-forming microorganisms of importance in foods, and few forums have been available for the exchange of information among investigators interested in the problems pertaining to these microorganisms. Therefore, the Committee recommends that, as a preliminary step in initiating research programs in the subjects noted here, an international workshop be convened for the exchange of information on recent developments and for discussion of the specific directions most appropriate for work in each subject.

It cannot be stated with certainty that the information derived from the studies outlined here will enable the rapid identification of easily exploitable points of susceptibility in the physiology of C. botulinum or other microorganisms. Even if such targets are identified, the development of agents that attack them and effectively inhibit undesired microorganisms will still be a long and possibly expensive process, as will be the toxicologic evaluation necessary to ensure that no adverse health effects would result from their use in foods. However, the Committee strongly recommends support for research into the physiology of spore-forming bacteria, particularly the spore-to-cell transition, in such microorganisms. Such research is far more likely than an empirical screening program to yield long-term improvement in the safety of cured meats and other food products.

Antioxidant Alternatives

Research on inhibition of lipid oxidation in cured products poses perhaps the most straightforward challenge in developing alternatives to nitrite. This subject is discussed fully in Chapter 6.

Alternatives for Sensory Effects

The most appropriate strategy for developing alternatives that confer sensory characteristics similar to those produced by nitrite will be determined largely by the results of research suggested in the Committee's first report (National Academy of Sciences, 1981) and Chapters 7 and 9 of this report. For example, it is necessary to know what contribution nitrite makes to flavor in different products, whether there is a dose-response relationship between nitrite addition and flavor, and, if so, what form it takes. Similarly, it is necessary to know how mutable are the putative consumer preferences for particular flavors and colors in products and how the ability to discriminate cured-meat flavor is distributed in the general population. Although it appears that nitrite makes a significant contribution to the flavor of ham and ham-based products, it is desirable to know how.

The answers to these questions should clarify the need for alternatives to nitrite in the production of sensory effects and help in defining the requirements of alternatives, if needed.

EVALUATION METHODS AND EXPERIMENTAL DESIGN

The Committee includes at this point a discussion of some considerations it believes are pertinent to the design of future studies to ensure that the maximal amount of useful data is provided and that such data are, as far as is possible, amenable to comparison among studies.

Antimicrobial Effects

One problem encountered by the Committee in its evaluation of alternatives to nitrite was in extracting comparable data from published reports on the efficacy of antimicrobial agents. This problem arose partly because nitrite has a multiplicity of antimicrobial effects. Many investigators have worked with *Clostridium botulinum*, but those more interested in putrefactive spoilage or not equipped to cope with pathogens like *C. botulinum* have often chosen to use *C. sporogenes*. A few have chosen to address the effects of nitrite on other microorganisms. Different end points have been used to measure the efficacy of nitrite or alternative agents. The end points used have included the percentage of inoculated packages that have become toxic, spoiled, or swollen at various times and the time to first appearance of swelling or toxicity in a set of packages. Selection of the best measure of antimicrobial efficacy is discussed below.

Investigators have also varied in their choice of system in which to test the antimicrobial activity of nitrite and alternatives to it. Systems have included test-tube or Petri-plate tests and the use of controlled, defined laboratory media and conditions; water slurries of meat heat-processed in tubes or bottles; meat products processed and generally vacuum-packaged in test tubes or cans; products made and packaged in pilot plants; and commercially made products. The classical test-tube approach offers more controlled conditions, a relatively short time to obtain data, ready availability of equipment, and lower research costs. The test-tube approach has yielded interesting results, but the practical applicability of some of the data to commercially made products is questionable. The antimicrobial activity of a particular compound in laboratory media may not be replicated in meat systems, particularly if the agent is lipophilic. For example, Robach *et al*. (1977) reported that butylated hydroxyanisole (BHA) at 50 mg/kg effectively inhibited the growth of *Vibrio parahaemolyticus* in trypticase soy broth, but 400 mg/kg was necessary for inhibition in a crabmeat homogenate. Kleindworth *et al*. (1979) reported a similar result on the effect of BHA on the growth of *Clostridium perfringens* in culture media with and without added lipid. Robach and Pierson (1978) reported inhibition of *C. botulinum* spores in culture media by propyl paraben at 25 mg/kg. But further work reported by Tanaka *et al*. (1978) indicated that parabens were not as effective in inhibiting *C. botulinum* in meat emulsions as in culture media. Work by Robach *et al*. (1981) indicated that glyceryl monolaurate was largely unavailable for antimicrobial action in the aqueous phase of food systems, because of its greater affinity for the lipids in such systems.

Whichever method is selected to test the antimicrobial efficacy of compounds, a compound must be thoroughly tested in model meat systems or in actual products before its feasibility as an alternative to nitrite can be determined.

The Committee did not feel it was justified in specifying one particular approach or testing method in preference to another, inasmuch as each provides useful information. However, outlined below are the issues that the Committee feels researchers and sponsors should consider in designing experiments to compare nitrite with possible alternatives.

Overall Objectives of Efficacy Evaluation. For assessment of antimicrobial agents as possible meat preservatives, experimentation must elucidate each agent's spectrum of antimicrobial action, its quantitative antimicrobial activity, the effects of product characteristics (e.g., pH, a_w, and brine concentration) that influence the activity, the effect of curing adjuncts (e.g., ascorbate and phosphates), and the effect of processing treatments that the agent would encounter if used. In cured meats, the microorganisms of primary concern are the spore-forming bacteria, particularly *C. botulinum*, in products heated sufficiently to destroy the more heat-sensitive vegetative cells of spoilage organisms and such foodborne pathogens as *Staphylococcus aureus* and salmonellae. However, other bacteria, molds, and yeasts are also of concern, particularly as spoilage agents. The ultimate environment in which the antimicrobial effect must be exerted is obviously the meat product, whose characteristics, as well as packaging and storage, may influence the activity of the agent. Testing can be done in laboratory media, meat slurries, or products; once the test system is chosen, a specific set of environmental conditions--pH, salt concentration, spore inoculum, etc.--must be defined and standardized. Thus, there are innumerable possible combinations of environmental factors in which the activity of agents against the spectrum of target organisms can be tested. In addition, the stability of the agent during production procedures, especially heat treatments and emulsifications, must be considered.

Information on the mechanism by which antimicrobial effects are brought about could also be sought; such information is often useful, for example, in identifying factors that might influence antimicrobial activity or possible toxic effects.

Test Organisms and Inoculation Procedures. *C. botulinum* is generally considered to be the species most significant for antimicrobial testing of potential meat preservatives, at least from a public-health point of view. The Committee concurs in this view, but suggests that the need to investigate efficacy against other pathogens and spoilage organisms, such as more heat-resistant clostridia like *C. sporogenes*, not be neglected. However, *C. botulinum* can serve as an example of the experimental design considerations that are also relevant to other microorganisms.

With *C. botulinum*, one must define the "type" (A, B, or E) or combination of types to be tested and select one or more strains representative of those occurring in nature, e.g., proteolytic or nonproteolytic. Spores or vegetative cells must be selected as the test inoculum, and the conditions of preparation of the inoculum (e.g., medium and incubation temperature) must be defined and standardized. The inocula must be stored under defined conditions that do not influence the results. The method of application to the food product under study must be defined. If a dry inoculum is used, so as not to change the water activity, the drying effects of the inoculum must be considered. If addition of the inoculum introduces water or culture medium, its influence on product water activity needs consideration. Standardization of inoculation procedures and documentation of the numbers of viable spores or cells inoculated are essential to facilitate comparisons of results among studies. Timing of inoculation must be defined, i.e., before processing, after the addition of the antimicrobial agent, or after the entire product has been processed. The inoculation time and approach will often be defined by the type of food product that is under consideration.

Test System. As noted above, with some types of antimicrobial agent--specifically highly lipophilic ones--the system in which antimicrobial efficacy is tested is extremely important. Activity in sterile laboratory media may differ significantly from behavior in commercial food materials processed under normal manufacturing conditions and is thus not a reliable guide to likely efficacy in commercial products. The Committee believes that agents must be tested in commercial products made under a variety of normal processing conditions before final decisions are made on the feasibility of those agents. Owing to the expense of such testing, it accepts that testing in model meat systems or in food materials produced under pilot-plant or laboratory conditions is a realistic compromise for preliminary or exploratory testing, especially when a highly contagious or dangerous inoculum is involved. Experimentation in laboratory media may be useful in determining mechanisms of action of agents or in providing initial insight into relevant but poorly understood basic phenomena, e.g., the effect of redox potential on microbial growth.

Measurement of Effects. The effect of an antimicrobial on the test inoculum must be measured in some way that is related to the likely adverse effects that the organism would produce if its proliferation were unchecked. In choosing an end point in studies of antimicrobial effects on pathogens, one should bear in mind the ultimate need to relate experimental observations to possible consequences for human health. Obviously, for foodborne intoxication, the presence of toxin in the product is the best measure of potential hazard; for infections, the number of organisms (e.g., salmonellae) is the best guide.

Currently available methods for measurement of botulinal toxin have been reviewed by Sugiyama (1980). They include mouse injection and an enzyme-linked immunosorbent assay (ELISA). The former method is expensive, laborious, and on occasion subject to unavoidable false positives from nonspecific toxicity due to product spoilage. The latter, in its present form, although less laborious, is not highly sensitive. Better methods for detecting *C. botulinum* toxin or growth would facilitate research on and testing of alternatives.

The average time to spoilage or toxicity of a particular percentage of packages is probably the most useful method of comparing alternatives, but the extremes of the ranges of these measures, e.g., first toxic package, are also important. They represent biologic variability, such as the clumping of spores or variation in resistance of organisms to antagonistic agents, and any one package that contains toxin can be lethal.

When the primary objective of the study is to determine antimicrobial efficacy against bacteria that cause spoilage, more general measures, such as observation of gas production or swelling of cans, may be acceptable as a means of reducing experimental costs. It should, however, be noted that the swelling of packages inoculated with *C. botulinum* does not necessarily indicate that they are toxic, and toxin can be present in unswollen packages (Sofos *et al.*, 1980).

Selection of appropriate controls and conditions is important for the accurate measurement of efficacy. Controls should be chosen to elicit the maximal amount of information obtainable from a particular study. For example, if a combination of agents is tested, the experimental design should enable the contribution of each agent and of additive or synergistic effects to be determined. Conditions should be relevant to those likely to be encountered by the product during handling, e.g., abuse at temperatures ranging from the minimum that permits growth of the test organisms to ambient temperatures (or above). Simulation of normal product handling should be considered, e.g., refrigerated storage followed by abuse, as well as abuse shortly after processing. Tests must be long enough to evaluate antimicrobial activity over the maximal period likely before product consumption.

Evaluation of Antimicrobial, Product, and Production-Process Variables. Consideration needs to be given to the form in which the antimicrobial is used in the test system (e.g., as free acid or salt) and to the form in which it is likely to be applied to the product during processing (e.g., as solid or in solution).

A host of factors largely determined by the nature of the product need to be selected, defined, or controlled. These include pH; solutes, including sodium chloride, sugar, and curing adjuncts; a_w; E_h; meat cut, type, age, species, and fat concentration; gaseous atmosphere; and processing temperatures and times. Such factors may vary within characteristic ranges for particular products. The efficacy of the test agent should be tested beyond the usual range of variation, to assess the consequences of failure to reach target values. The degree of product contamination, simulated by the inoculum of test microorganisms, should be regarded as a test variable, and different concentrations of inoculum should be used.

Data Collection and Analysis. Careful attention to all the foregoing considerations may be rendered almost useless, if prior thought is not also given to collection, organization, and analysis of the data generated. Methods could be coordinated between large studies, to facilitate interpretation of the meaning of results for "real-world" situations and to enable comparisons of antimicrobial efficacy to be made easily.

Discussion. There is a need to test the antimicrobial efficacy of alternatives to nitrite against the wide variety of potential pathogens and spoilage organisms in cured meats, in all products in which their use is contemplated, under conditions that simulate those likely under normal and inadequate product processing and handling, and with due regard to the variations in characteristics of the product, such as pH and water activity. Specifications of protocols for testing have not been recommended by the Committee, for a variety of reasons. Investigators will have particular interests in examining alternatives, and generalized protocols may not be appropriate to all such situations. Design of protocols to cover all or even most situations would be a major undertaking that was beyond the Committee's charge. Useful information can be generated by well-designed experiments conducted on a small scale or by novel approaches that might be denied support if they did not conform to a particular protocol intended to describe ideal, comprehensive studies.

The Committee does, however, strongly recommend that preservative testing be done with food materials as similar as possible to the products in which ultimate use is intended.

Antioxidant Effects

The methods available for testing the efficacy of alternatives to nitrite for inhibition of lipid oxidation are discussed in Chapter 6. Of these methods, it appears that thiobarbituric acid values, used where possible in conjunction with sensory evaluation, would be the most useful for comparison.

Testing for efficacy of inhibition of lipid oxidation should be conducted in such a manner as to simulate the storage and handling that products receive during distribution and in consumers' possession.

Sensory Effects

Alternatives to nitrite can be evaluated to answer two questions: Do consumers perceive the product made with the alternative to be identical with the conventional one? If not identical, is it as acceptable as the conventional one?

The Committee believes that satisfying the second (hedonic) criterion justifies classification of an alternative as promising. However, it also believes that sensory evaluation of specific attributes by analytic panels can provide useful information and should be included in studies of alternatives.

Evaluations should, as far as possible, use experimental designs that allow the magnitude of sensory differences among products to be determined and that show whether the ability to perceive or discriminate an attribute conferred by an alternative is limited to particular people.

Because of the cuing effect of color on flavor (DuBose _et al._, 1981), sensory evaluations of flavor should be conducted in color-masking situations.

Most sensory studies with nitrite have been conducted on products shortly after processing. There is a need to conduct further sensory studies on nitrite and alternatives under protocols that elicit information on sensory effects over the normal storage life of the products in question.

Rankings of product acceptability as determined by laboratory preference studies do not necessarily predict general consumer acceptance. The reasons for this may be regional or ethnic differences in preference or variations in sensory ability. Hence, large-scale consumer tests should be conducted on alternatives that pass initial hedonic screening with analytic panels.

CONCLUSIONS

To duplicate most closely the important antimicrobial effects of nitrite, research intended to develop alternatives should focus on agents or techniques that prevent or inhibit the development of bacterial spores into vegetative cells, particularly spores of *Clostridium botulinum*.

The rational long-term approach to the search for such antimicrobial alternatives would be to capitalize on knowledge of biochemical differences between humans and the undesired microorganisms and to design and develop agents that inhibit metabolic processes that are essential to such microorganisms, but are absent in humans. Stages in the spore-to-cell transition probably offer a number of opportunities for such intervention. However, too little is known about these processes to justify focusing on a single target in a program designed to develop agents that inhibit this transition.

Research is needed on the physiology of spore-forming microorganisms, particularly on spore dormancy, injury, activation, germination, and outgrowth and factors that affect these characteristics. Such research would facilitate identification of peculiarities in metabolic processes essential to the spore-to-cell transition, and that identification might profitably be exploited to inhibit it. The research program necessary to obtain this knowledge and the subsequent program to design and develop selective antimicrobial agents are likely to be laborious and expensive. There is no guarantee that they will lead to the rapid development of practicable alternatives to nitrite. However, the Committee regards this approach as a necessary long-term investment in developing alternatives to nitrite and as one that might contribute to increasing the safety and freedom from spoilage of other types of product in which spore-forming microorganisms pose problems.

Efficacy testing of potential antimicrobial alternatives must ultimately be conducted in the products in which use is intended, although some testing in model meat systems may be justified for preliminary or exploratory investigations. A large number of other factors--e.g., variation in product characteristics, method chosen to assess efficacy, and timing and duration of simulated temperature abuse--need to be considered in designing experiments. However, the Committee concluded that specifying ideal protocols was neither practicable nor desirable.

Alternatives to nitrite for antimicrobial and antioxidant effects need to be subjected to hedonic (acceptability) testing and sensory evaluation, with ratio-scaling methods to elicit a maximum of useful information. However, long-term strategies for developing alternatives to nitrite for the production of sensory effects (color and flavor) can be chosen only after completion of studies to define the need for alternatives for these effects, i.e., studies of the strength of consumer preferences for cured-meat color, of the magnitude of nitrite's flavor contribution in various products, and of the mechanism whereby this is achieved.

Testing for antioxidant effects should be conducted under conditions similar to those under which the product is distributed and handled by consumers.

RECOMMENDATIONS

As part of a long-term strategy to develop antimicrobial alternatives to nitrite, the Committee recommends that an international workshop be convened to review new information on the the physiology of spore-forming microorganisms (with emphasis on anaerobes) and, on the basis of this information, to suggest potentially profitable directions for future research to develop methods for their control.

The Committee recommends that investigations of spore dormancy, injury, activation, germination, and particularly outgrowth be conducted, preferably with *Clostridium botulinum*. The mechanism of action of known inhibitors of spore-forming microorganisms, preferably nitrite, should also be investigated.

Research directed at developing reliable, rapid, simple techniques to detect *C. botulinum* growth or toxin production should also be undertaken.

Recommendations aimed at better defining the need for alternatives to nitrite for color and flavor effects are included in Chapters 7 and 9.

REFERENCES

Ando, Y. 1973. [In Japanese; English Summary.] Studies on germination of spores of clostridial species capable of causing food poisoning. I. Factors affecting the germination of spores of *Clostridium botulinum* type A in a chemically defined medium. Shokuhin Eiseigaku Zasshi 14:457-461.

Ando, Y. 1974. Alpha-hydroxy acids as co-germinants for some clostridial spores. Jpn. J. Microbiol. 18:100-102.

Benedict, R. C. 1980. Biochemical basis for nitrite-inhibition of *Clostridium botulinum* in cured meat. J. Food Prot. 43:877-891.

DuBose, C. N., A. V. Cardello, and D. Maller. 1981. Factors affecting the acceptability of low-nitrite smoked cured ham. J. Food Sci. 46:461-463.

Foegeding, P. M., and F. F. Busta. 1981. Bacterial spore injury: An update. J. Food Prot. 44:776-786.

Kleindworth, K. J., P. M. Davidson, C. J. Brekke, and A. L. Branen. 1979. Inhibition of *Clostridium perfringens* by butylated hydroxyanisole. J. Food Sci. 46:566-567.

National Academy of Sciences. 1981. The Health Effects of Nitrate, Nitrite, and *N*-Nitroso Compounds. National Academy Press, Washington, D.C. 544 pp.

Robach, M. C., and M. D. Pierson. 1978. Influence of para-hydroxybenzoic acid and esters on the growth and toxin production of *Clostridium botulinum* 10755A. J. Food Sci. 43:787-792.

Robach, M. C., L. A. Smoot, and M. D. Pierson. 1977. Inhibition of *Vibrio parahaemolyticus* 04:K11 by butylated hydroxyanisole. J. Food Prot. 40:549-551.

Robach, M. C., C. S. Hickey, and E. C. To. 1981. Comparison of antimicrobial actions of monolaurin and sorbic acid. J. Food Safety 3:89-98.

Sakaguchi, G. 1979. Botulism. Pp. 389-442 in H. Riemann and F. L. Bryan, eds. Food-borne Infections and Intoxications, 2nd Ed. Academic Press, New York. 748 pp.

Smith, L. DS. 1977. Botulism: The Organism, Its Toxins, The Disease. Charles C Thomas, Springfield, Illinois. 236 pp.

Smoot, L. A., and M. D. Pierson. In press. Inhibition and control of bacterial spore germination. J. Food Prot.

Sofos, J. N., F. F. Busta, K. Bhothipaksa, C. E. Allen, M. C. Robach, and M. W. Paquette. 1980. Effects of various concentrations of sodium nitrite and potassium sorbate on _Clostridium botulinum_ toxin production in commercially prepared bacon. J. Food Sci. 45:1285-1292.

Sugiyama, H. 1980. _Clostridium botulinum_ neurotoxin. Microb. Rev. 44:419-448.

Tanaka, N., J. G. Cerveny, N. J. Worley, E. W. Wall, and J. M. Goepfert. 1978. Effect of nitrite, sorbate, polyphosphates and parabens on toxin production by _Clostridium botulinum_ in wieners. Pp. 354-358 in Annual Report of the Food Research Institute, University of Wisconsin, Madison, Wisconsin.

APPENDIX A

TESTING OF COMPOUNDS IN LABORATORY MEDIA, MEAT SLURRIES OR MODEL SYSTEMS FOR ANTIMICROBIAL ACTIVITY: A SELECTED BIBLIOGRAPHY

Chang, N. C., and A. L. Branen. 1975. Antimicrobial effect of butylated hydroxyanisole. J. Food Sci. 40:369-351.

DeWitt, J. C., S. Notermans, N. Gorin, and E. H. Kampelmacher. 1979. Effect of garlic oil or onion oil on toxin production by *Clostridium botulinum* in meat slurry. J. Food Prot. 42:222-224.

Dymicky, M., and C. N. Huhtanen. 1979. Inhibition of *Clostridium botulinum* by p-hydroxybenzoic acid n-alkyl esters. Antimicrob. Agents Chemother. 15:798-801.

Dymicky, M., C. N. Huhtanen, and A. E. Wasserman. 1977. Inhibition of *Clostridium botulinum* by 5-nitrothiazoles. Antimicrob. Agents Chemother. 12:353-356.

Gibson, A. M., T. A. Roberts, and A. Robinson. In press. Factors controlling the growth of *Clostridium botulinum* types A and B in pasteurized cured meats. IV. Effect of pig breed, cut and batch of pork. J. Food Technol.

Huhtanen, C. N. 1979. Bile acid inhibition of *Clostridium botulinum*. Appl. Environ. Microbiol. 38:216-218.

Huhtanen, C. N. 1980. Inhibition of *Clostridium botulinum* by spice extracts and aliphatic alcohols. J. Food Prot. 43:195-197.

Huhtanen, C. N., and T. J. Micich. 1978. Inhibition of *Clostridium botulinum* by aliphatic amines and long chain aliphatic amino-diamides. J. Am. Oil Chem. Soc. 55:854-855.

Incze, K., J. Forkas, V. Miháyl, and E. Zukal. 1974. Antibacterial effect of cysteine-nitrosothiol and possible precursors thereof. Appl. Microb. 27:202-205.

Kleindworth, K. J., P. M. Davidson, C. J. Brekee, and A. L. Branen. 1979. Inhibition of *Clostridium perfringens* by butylated hydroxyanisole. J. Food Sci. 46:564-567.

Pierson, M. D., L. A. Smoot, and K. R. Van Tassell. 1980. Inhibition of *Salmonella typhimurium* and *Staphylococcus aureus* by butylated hydroxyanisole and the alkyl ester of p-hydroxybenzoic acid. J. Food Prot. 45:191-194.

Rhodes A. C., and B. Jarvis. 1976. A pork slurry system for studying inhibition of *Clostridium botulinum* by curing salts. J. Food Technol. 11:13-23.

Robach, M. C., and M. D. Pierson. 1978. Influence of parahydroxybenzoic acid esters on the growth and toxin production of *Clostridium botulinum* 10755A. J. Food Sci. 43:787-792.

Robach, M. C., F. J. Ivey, and C. S. Hickey. 1978. System for evaluating clostridial inhibition in cured meat products. Appl. Environ. Microbiol. 36:210-211.

Robach, M. C., C. S. Hickey, and E. C. To. 1981. Comparison of antimicrobial actions of monolaurin and sorbic acid. J. Food Safety 3:89-98.

Roberts, T. A., A. M. Gibson, and A. Robinson. 1981. Factors controlling the growth of *Clostridium botulinum* types A and B in pasteurized cured meats. I. Growth in pork slurries prepared from "low" pH meat (pH range 5.5-6.3). J. Food Technol. 16:239-266.

Roberts, T. A., A. M. Gibson, and A. Robinson. 1981. Factors controlling the growth of *Clostridium botulinum* types A and B in pasteurized cured meats. II. Growth in pork slurries prepared from "high" pH meat (pH range 6.3-6.8). J. Food Technol. 16:267-281.

Roberts, T. A., A. M. Gibson, and A. Robinson. 1981. Prediction of toxin production by *Clostridium botulinum* in pasteurized pork slurry. J. Food Technol. 16:337-355.

Roberts, T. A., A. M. Gibson, and A. Robinson. In press. Factors controlling the growth of *Clostridium botulinum* types A and B in pasteurized cured meats. III. Effect of potassium sorbate. J. Food Technol.

Scott, V. N., and S. L. Taylor. 1981. Temperature pH and spore load effects on the ability of nisin to prevent the outgrowth of *Clostridium botulinum* spores. J. Food Sci. 46:121-126.

Scott, V. N., and S. L. Taylor. 1981. The effect of nisin on the outgrowth of *Clostridium botulinum* spores. J. Food Sci. 46:117-120, 126.

Smoot, L. A., and M. D. Pierson. 1979. Effect of oxidation-reduction potential on the outgrowth and chemical inhibition of *Clostridium botulinum* 10755A spores. J. Food Sci. 44:700-704.

Stern, N. J., L. A. Smoot, and M. D. Pierson. 1979. Inhibition of *Staphylococcus aureus* growth by combinations of butylated hydroxyaniosole, sodium chloride, and pH. J. Food Sci. 44:710-712.

APPENDIX B

SOME RECENT ADVANCES IN THE DESIGN OF INHIBITORS OF CELLULAR GROWTH

By

Seymour S. Cohen
Department of Pharmacological Sciences
State University of New York, Stony Brook

In attempting to inhibit the growth of bacteria in food with chemical agents that may replace nitrate or nitrite, we are interested in substances that will not poison or inhibit the feeding mammal. The problem is formally similar in some respects to the problem of developing a selective chemotherapeutic agent to arrest growth or development of an infecting organism, e.g., a sulfa drug or a penicillin. However, numerous other considerations, of which cost is only one, make an already difficult problem even more difficult. It is appropriate to ask whether the design of drugs for selective chemotherapy has advanced sufficiently to permit any hope in our problem-solving. Obviously, the existence of sulfa drugs and antibiotics and their improvements do point to the accomplishments of a refined and increasingly sophisticated empiricism. In addition to the screening and modifying efforts that have yielded these successes, modern science and technology have made advances in recent years that permit a general approach to the design of discriminating inhibitory agents (Cohen, 1977, 1979, 1981). In considering the matter of a replacement for nitrite, we shall concentrate on this avenue of search for an antibacterial drug.

Three major components of our advances, taken together, can provide a strategy focusing on essential proteins of microorganisms as our selected targets. These advances can be summarized in a historical framework as follows:

- It was established from the mid-1930s to the end of the 1950s that prokaryotic organisms have or determine specific structures and metabolic systems essential for their survival or multiplication. Such systems should be specifically inhibitable, as in the action of penicillin on synthesis of bacterial cell walls or of some other antibiotics on bacterial ribosomes.

- The molecular biology of the 1960s and 1970s established that the genetic determinants and therefore the proteins of microorganisms are different from the DNA and proteins of the eukaryotic cells of animals. These major differences in sequence and conformation exist often for proteins and enzymes that have similar functions.

• In the 1970s, it became possible to isolate proteins far more easily from relatively small amounts of biologic materials. The characterization of proteins with respect to amino acid sequence, conformation, surface structure, and reactivities has been developed in a manner that should permit chemists to devise specific tight-binding inhibitors. The synthesis and properties of such inhibitors are better understood, as are some of the problems involved in ensuring the penetration of the inhibitors.

GENERAL CONSIDERATIONS

We are interested in affecting proteins, rather than nucleic acids, because it seems more feasible to prepare specific inhibitors that may not be mutagenic. One reason for our concern with nitrite is its role in generating mutagenic nitrosamines. It would be reasonable to focus on compounds that would not be expected to enter nucleic acids or otherwise distort their structure.

It may be argued that the chances of finding an inexpensive substitute for nitrite are so small that there is no point in seeking it through the process of defining selectively inhibitable essential proteins. That may prove to be so, but it is not self-evident, inasmuch as the effectiveness of many other low-molecular-weight inhibitors is known. In addition to numerous substances that trap metals required for catalytic functions, we may point to a gas like ethylene, which is used as a fruit-ripening agent and also inactivates cytochrome P_{450} (Ortiz de Montellano and Mico, 1980), or phosphonoformate, the specific, relatively nontoxic competitive analogue of pyrophosphate, which inhibits the nucleate polymerases induced by some DNA viruses and RNA retroviruses (Eriksson *et al.*, 1980).

The comparative biochemistry illustrative of the possibilities of selective chemotherapy is elaborated by Rogers in Appendix C. But it must be asked whether the desired inhibitor is intended to act on a broad spectrum of bacteria, on a limited spectrum of anaerobes, or merely on a toxigenic *Clostridium botulinum*. The biologic breadth of the target(s) must be defined before the biochemical and chemical effort is planned. Even if the cellular target is stated to be exclusively the germination of spores of *C. botulinum*, it appears clear that the possible structural and catalytic targets are ill-defined, in that the data on the physiology of germination of the organism are woefully sparse. In brief, it is necessary to define the essentiality of any protein target to the survival or development of the organism, and there does not appear to have been enough work on the physiology of *C. botulinum* to identify an essential protein target.

The study of the physiology of *C. botulinum* will require attention to anaerobic techniques that are known, although infrequently practiced. Toxin production is understood to be determined by a gene of a lysogenic phage. Therefore, much of the physiology of the bacterium can be explored in nontoxigenic strains. Proteins essential to toxin production can be defined through defining the phage gene(s) determining this function. Clearly, the project needs the active participation of microbiologists competent in many specialties.

PROTEIN PRODUCTION AND ISOLATION

For purposes of isolating and analyzing the protein essential to spore germination or toxin production, it is important to develop organisms that maximize synthesis of the particular protein. Inhibitor-resistant organisms or regulatory mutants may be useful overproducers, or the clostridial or phage structural genes for the desired protein can be cloned by modern techniques of genetic engineering in some bacterial host to maximize synthesis of the particular protein.

Some 15-20 years ago, the isolation of proteins to homogeneity was an enormous, tedious, and difficult task that involved large amounts of starting material. Modern techniques of sizing with gels and adsorption chromatography to manipulate charge and specificity, as in affinity chromatography, have made spectacular simplifications in time, amount of material, and degree of purification required. Major improvements in gel electrophoresis have also contributed to the isolation and characterization of the various proteins, as well as to the analysis of the biologic heterogeneity of numerous enzymes.

For many proteins, isolation to homogeneity has become a reasonable task for an experienced laboratory. For example, the purification to homogeneity of a glutamate dehydrogenase of *Trypanosoma cruzi* has been accomplished with a 1-g batch of the organism, with determination of the subunit molecular weight of 64,000 in the tetrameric protein and of some kinetic properties of the pure enzyme.

PROTEIN CHARACTERIZATION

Recent experience with the isolation and characterization of interferon has compelled the development of improved microtechniques. New protein sequenators (Hunkapiller and Hood, 1980) have now permitted the analysis of the 13 amino-terminal amino acids in 1- to 2-μg quantities of interferon, an increase in sensitivity by a factor of 10^4 over a method described by Edman in 1967. Molecules of over 1,000 amino acids have been sequenced somewhat slowly, but genetic recombinant techniques and sequence analysis for DNA have permitted amino-acid sequences to be determined far more rapidly. Thus, knowledge of the sequences of nucleotides in the thymidine kinase gene of herpes simplex virus type 1 (Wagner *et al.*, 1981) facilitates the acquisition of knowledge of the protein structure of this virus-induced enzyme before its characterization as a protein is complete. A similar knowledge of the sequence of nucleotides in poliovirus RNA has permitted the definition of the amino-acid sequence of a poliovirus-induced protease. A knowledge of amino-acid sequence can provide an initial approach to the problem of protein conformation, and eventually to inhibitor-sensitive sites.

Many workers have studied enzymes almost entirely in terms of their active sites and known substrates, products, and postulated reaction intermediates. Much synthetic work leading to carefully analyzed structure-activity relationships has led to the informed synthesis of excellent inhibitors of enzymes. A recent success in this direction of rational design has been the development of Captopril, an inhibitor of the angiotensin-converting enzyme, which was assisted by a comparison of this enzyme with the known active site of carboxypeptidase A (Cushman and Ondetti, 1980). Such an approach might be helpful in designing a specific inhibitor of the poliovirus-induced protease, which is essential to the specific processing of the giant protein elaborated by translation of poliovirus RNA.

However, we shall suppose that knowledgeable design of an inhibitor of protein function will require the detailed study of topology and conformation of the protein, i.e., its three-dimensional structure. Even in the absence of such knowledge, one may approach analysis of the active site of an enzyme through study of the volume occupied by a series of substrates and inhibitory, conformationally rigid analogues (Sufrin *et al.*, 1981). The volumes can be determined from x-ray crystallographic analysis of the inhibitory analogues. A similar analysis of noninhibitory compounds can be instructive as to the portion of the molecule that cannot fit within the enzyme site, i.e., the enzyme-essential volume.

More than 150 proteins have been crystallized and analyzed at a high level of resolution. The improvement of computerized analytic techniques, including computer imaging, has made the analysis of the fine detail of shape, sequence, and topology available for the analysis of substrate and inhibitor interaction. Indeed, in many instances the cocrystallization of an inhibitor-protein complex has been essential to the crystallization and analysis of the protein. Some advanced laboratories have attempted to use such data to assist in drug design in the last 5 yr, and some successes in these types of "molecular modeling" have been recorded, as in a recent report from Merck (Gund *et al.*, 1980). That laboratory and that of Burroughs Wellcome have been attempting to use these techniques to design improved inhibitors of dihydrofolate reductase (DHFR) of *E. coli*, an enzyme readily purified by affinity chromatography on methotrexate-containing adsorbates and readily crystallized with various inhibitors at the active site.

Most recently, the Merck group has reported patenting such a new inhibitor designed with this type of crystallographic procedure (Poe et al., 1981). The new inhibitor is stated to "form a hydrogen bond from the sulfonamide oxygen to the peptide NH of LEU 23, form a hydrophobic bond to the sidechain of TRP 22, and form a hydrogen bond to the peptide carbonyl of ALA 19, when the pyrimidine ring is positioned identically to the pyrimidine ring of methotrexate on DHFR." It is not clear that this inhibitor, or one made by Burroughs Wellcome, will serve as a super trimethoprim in control of gram-negative bacterial infections in the human urinary tract. Nevertheless, the power of the process in inhibitor design exemplifies some of the recent achievements in structural chemistry.

SOME ADDITIONAL METHODS

Allusions may be made to the work of Atassi (1978) in the development of peptides complementary to antigenic binding sites. In studies on lysozyme, three major antigenic sites, accounting for 85-90% of antilysozyme antibody in animals immunized to the enzyme, were defined. The component amino acids of each site proved to be noncontiguous in the rigid polypeptide, but formed a sequence at the surface of the protein molecule, as determined by x-ray crystallography, immunochemical analysis, and "surface simulation." Sequences simulating the sites were then synthesized and shown to combine with antibody. Hexapeptides complementary to two of these sites were shown to bind specifically to lysozyme.

To fix a complementary structure as small as a hexapeptide to a protein surface, it may be necessary to develop an irreversible binding of peptide to the protein. Several kinds of almost irreversible inhibitors are now recognized, including those which form a covalent bond with a reactive residue of a protein during recognition steps and those which form a tight bond (either covalent or noncovalent) as a result of the transformation of the substrate during the catalytic steps of the enzymatic process.

Finally, reference should be made to studies attempting to control the molecular pathology of sickle-cell anemia by inhibiting the linear polymerization of deoxygenated sickle-cell hemoglobin (hemoglobin S) and the consequent deformation of the erythrocytes. In the genetically determined disease, valine replaces glutamate at a unique position in the β chain. The interaction of α and β subunits to produce intracellular gelation are controlled by the surface topology of the subunits, and replacements of some amino acids in the α subunits markedly inhibit gelation (Benesch et al., 1979). Modification of intrasubunit interaction and gelation has been actively pursued. Some increase in the solubilization of deoxygenated hemoglobin S has been observed with peptides containing phenylalanine, although it has been asked whether the hydrophobic bonds that facilitate polymerization can be easily affected by simple peptides. Nevertheless, the problem of affecting the spontaneous assembly of protein subunits is a general problem of virus multiplication and many aspects of microbial growth and development.

CONCLUSIONS

The strides made in biology and chemistry in recent years have permitted the formulation of a novel generalizable strategy of drug design for the purpose of affecting protein structure and function in numerous aspects of biologic activity. Such a strategy narrows the target to a specific essential protein, and in so doing may be inapplicable to the requirements of nitrite replacement. However, it has not yet been stated that nitrite is needed for inhibiting many functions and many proteins. If nitrite proved to have but a single major function--e.g., inhibiting ferredoxins of many species of microorganisms--one could still examine the problem in the terms outlined above. Furthermore, if it were clear that the main function of nitrite were indeed to inhibit the germination of spores of *Clostridium botulinum* and thereby to inhibit toxin production, the strategy outlined here would be feasible, even though it calls for a commitment to expensive modern multidisciplinary science and technology. In short, a precondition for decision-making in developing our approach is greater clarity of the biologic and chemical roles of nitrite in restraining microbial growth in various foods.

REFERENCES

Atassi, M. Z. 1978. The precise and entire antigenic structure of lysozyme: Implications of surface-simulation synthesis and the molecular features of protein antigenic sites. Adv. in Exp. Med. Biol. 98:41-99.

Benesch, R. E., S. Kwong, R. Edalji, and R. Benesch. 1979. Chain mutations with opposite effects on the gelation of hemoglobin S. J. Biol. Chem. 254:8169-8172.

Cohen, S. S. 1977. A strategy for the chemotherapy of infectious disease. Science 197:431-432.

Cohen, S. S. 1979. Comparative biochemistry and drug design for infectious disease. Science 205:964-971.

Cohen, S. S. 1981. Modern science and the design of drugs for infectious disease. Interciencia 6:146-155.

Cushman, D. W., and M. A. Ondetti. 1980. Inhibitors of angiotensin-converting enzyme. Pp. 41-104 in G. P. Ellis, and G. B. West, eds. Progress in Medicinal Chemistry: 17. Elsevier/North-Holland Biomedical Press, New York, New York.

Eriksson, B., A. Larsson, E. Helgstrand, N. Johansson, and B. Oberg. 1980. Pyrophosphate analogues as inhibitors of herpes simplex virus type I DNA polymerase. Biochim. Biophys. Acta 607:53-64.

Gund, P., J. D. Audose, J. B. Rhodes, and G. M. Smith. 1980. Three-dimensional molecular modeling and drug design. Science 208:1425-1431.

Hunkapiller, M. W., and L. E. Hood. 1980. New protein sequenator with increased sensitivity. Science 207:523-525.

Ortiz de Montellano, P. R., and B. A. Mico. 1980. Destruction of cytochrome P-450 by ethylene and other olefins. Mol. Pharmacol. 18:128-135.

Poe, M., J. P. Springer, and K. Hoogsteen. 1981. Comparison of the conformation of 2,4-diamino-5[4-methoxy-3-(2-sulfanilamidoethoxy) benzyl]pyrimidine in a crystal and on *Escherichia coli* dihydrofolate reductase (DHFR). Fed. Proc., Fed. Am. Soc. Exp. Biol. 40:1837.

Sufrin, J. R., D. A. Dunn, and G. R. Marshall. 1981. Steric mapping of the L-methionine binding site of ATP: L-methionine S-adenosyltransferase. Mol. Pharmacol. 19:307-313.

Wagner, M. J., J. A. Sharp, and W. C. Summers. 1981. Nucleotide sequence of the thymidine kinase gene of herpes simplex virus type I. Proc. Natl. Acad. Sci. 78:1441-1445.

APPENDIX C

APPROACHES TO DESIGN OF AGENTS FOR CONTROL OF *CLOSTRIDIUM BOTULINUM* IN FOODSTUFFS

By

Edward F. Rogers
Merck, Sharp and Dohme, Rahway, New Jersey

Discovery of a suitable antibacterial replacement for nitrite is a formidable assignment. The replacement must be stable during food storage and processing, must be extremely well tolerated and without effect on taste or odor, must have negligible resistance liability, and, allowing a premium for greater safety, must be commensurate in cost effectiveness.

As stated by Cohen (Appendix B), fundamental biochemical research on *Clostridium* spp. is necessary so that enzymes that are appropriate targets for inhibitor design may be identified. Suitable enzymes should be specific because the likelihood of toxicity to other species is thereby reduced; should be targets for which inhibitor design appears feasible (e.g., substrates to be simulated must not be too complex in structure); and should be targets based on a confirmed strategy (i.e., potent enzyme inhibitors produce the desired *in vivo* results, assuming solution of delivery problems). It must be emphasized that apparently plausible strategies often fail because of unrecognized negative factors, so prompt verification of a strategy is desirable.

Recent work on the mechanism of action of nitrite and *S*-nitroso compounds is impressive, but it is difficult to see how to use it. If there are sites of nitrite action besides the spore membrane, as appears probable (the failure to encounter nitrite-resistant clostridia argues for multiple sites), one may speculate on what they are. Inasmuch as there is evidence of iron reversal of nitrite action (Tompkin *et al*., 1979), perhaps products formed from nitrite affect iron utilization. Hence, iron-binding compounds, such as 2,3-dihydroxybenzoic acid and the siderophores shown in Figure C-1, may deserve attention.

One attractive inhibitor target is the dependence of *C. botulinum* on arginine as a source of citrulline, which in turn reacts with inorganic phosphate to form carbamyl phosphate. Typical antimetabolites for use in exploration of this subject are robenidine (Figure C-2: A), a coccidiostat that is a potent arginase inhibitor, and

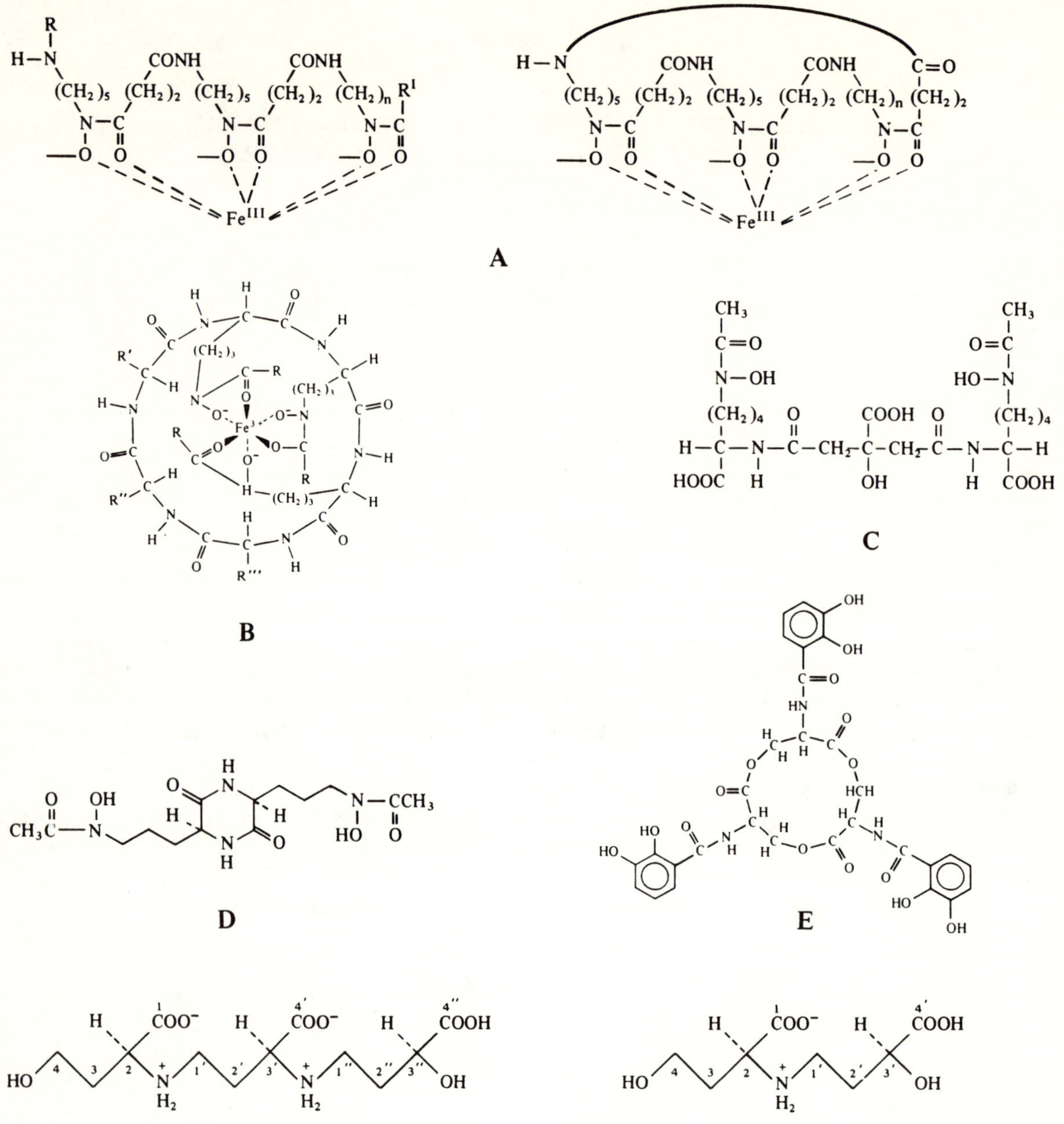

FIGURE C-1. Structural formulas of representative siderophores. A, linear and cyclic ferrioxamines. When it is linear and R = H, R' = CH_3, the compound is ferrioxamine B; when it is cyclic and n = 5, the compound is ferrioxamine E. B, ferrichromes. For R' = R" = R''' = H and R = CH_3, the compound is ferrichrome itself. C, aerobactin; D, rhodotorulic acid; E, enterobactin; F, avenic acid A; G, avenic acid B. From Raymond and Carrano, 1979, and Fushiya et al., 1980.

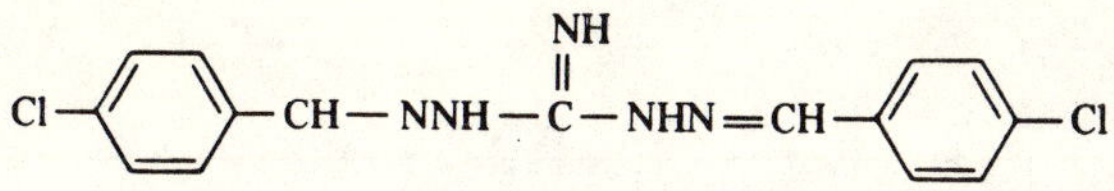

A

$HOOCCH(NH_2)CH_2CH_2CH_2NHCOCH_2PO_3H_2$

B

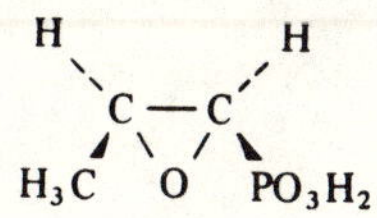

C

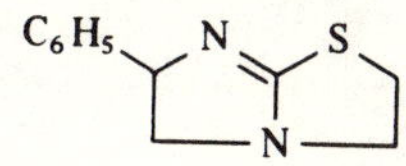

D

$NH(PO_3H_2)_2$

E

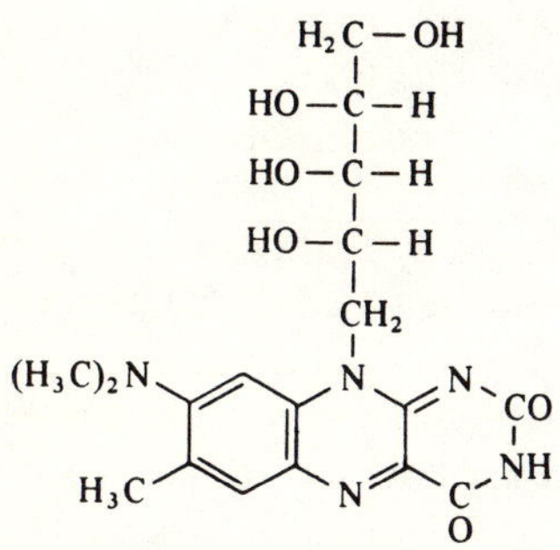

G

$CH_3COP(O)(OCH_3)ONa$

F

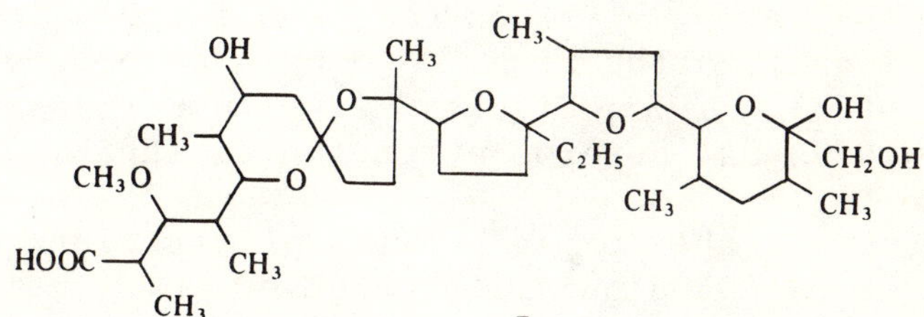

I

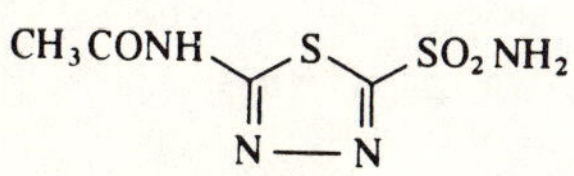

H

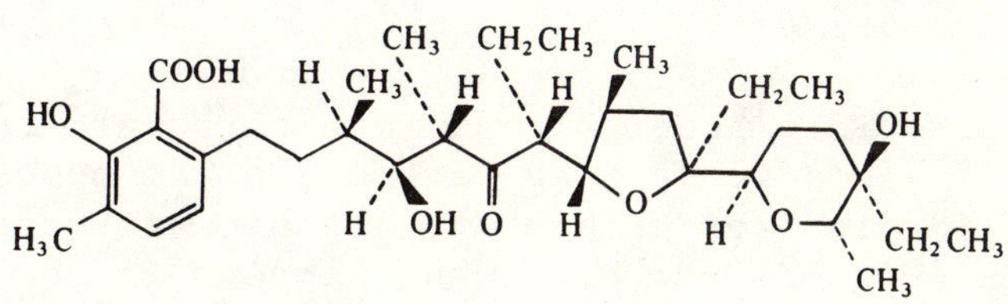

J

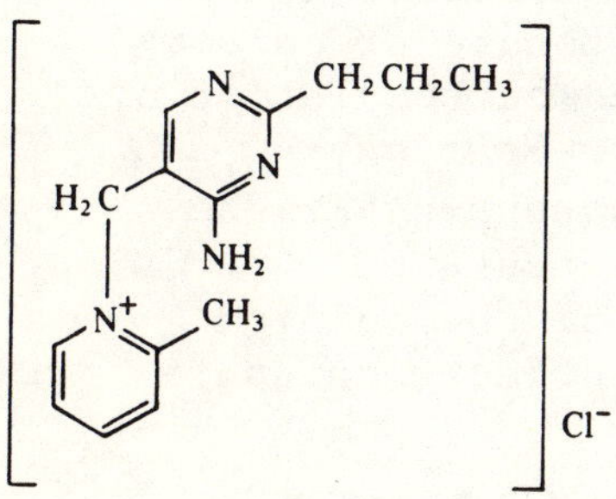

K

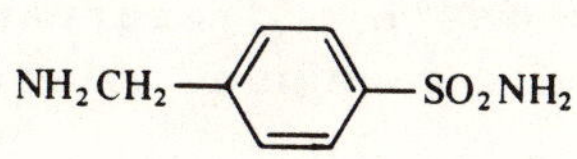

L

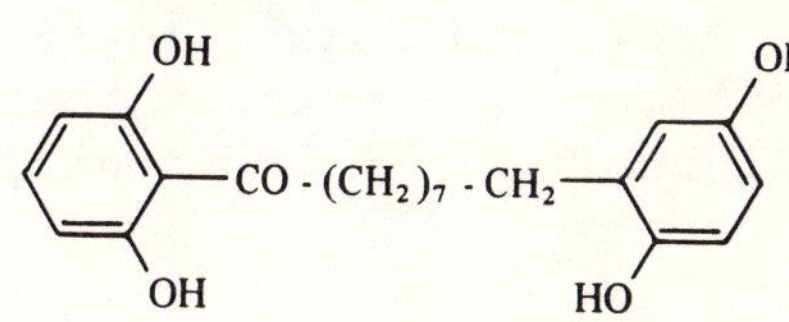

M

FIGURE C-2. Compounds of interest as possible inhibitors of _C. botulinum_. A, robenidine; B, N^5-phosphonoacetylornithine; C, phosphonomycin; D, tetramisole; E, imidodiphosphate; F, sodium methyl acetylphosphonate; G, roseoflavin; H, diamox; I, monensin; J, lasalocid; K, amprolium; L, marfanil; M. 2.6-dihydroxy-9-(2,5-dihydroxyphenyl)nonanophenone.

N^5-phosphonoacetylornithine (Figure C-2: B), a strong inhibitor of the second reaction. For further probes of anaerobic energy metabolism, recently reviewed by Gottschalk and Andreesen (1979), one may suggest tests of the antibiotic phosphonomycin (Figure C-2: C), a phosphoenolpyruvate antagonist in biosynthesis of muramic acid; tetramisole (Figure C-2: D), a fumarate reductase inhibitor; imidodiphosphate (Figure C-2: E), a pyrophosphate antagonist; and sodium methyl acetylphosphonate (Figure C-2: F), a pyruvate analogue (Kluger and Pike, 1977).

Figure C-3 outlines target areas in anaerobic energy metabolism, the pyruvate: ferredoxin oxidoreductase-ferredoxin hydrogenase and the pyruvate-formate lyase systems. Figure C-4 is a "three-dimensional" representation of the structure of oxidized *Clostridium* MP flavodoxin, in which the location of the flavin moiety is indicated. According to Walsh (1979), "it would appear that the accessible site for potential redox reaction is the dimethylbenzenoid end of the isoalloxazine ring." In this case, the antibiotic roseoflavin (Kasai *et al.*, 1975; Figure C-2: G) and related compounds (Graham *et al.*, 1977) may deserve attention as inhibitors.

If the complex series of reactions for fermentation of lysine to acetate and butyrate observed in *C. sticklandii* and *Clostridium* SB4 is relevant to *C. botulinum*, there is a plethora of targets. Perhaps lysine racemase would be most specific. It is hoped that 3-fluoro-D-lysine would perform as well as 3-fluoro-D-alanine does against alanine racemase.

Clostridial spores require bicarbonate for germination (Ando, 1973). Is conversion to carbon dioxide necessary? There may be justification for trials of carbonic anhydrase inhibitors, such as diamox (Figure C-2: H). What happens to the bicarbonate or carbon dioxide? One is reminded of the vitamin K-mediated carboxylation of protein glutamate residues to form calcium-binding γ-carboxyglutamate moieties. Calcium ionophores, such as monensin (Figure C-2: I) and lasalocid (Figure C-2: J), may deserve tests on this basis or for another reason--possible disturbance of the calcium dipicolinate concentration and consequent adverse effects on spore integrity.

Clostridia may have unusual thiamine biochemistry. Growth-media formulas seem to place special emphasis on this vitamin; and, curiously, many strains of *C. botulinum* excrete thiaminase, a rarely encountered enzyme (Hayashi *et al.*, 1973). I have developed a class of coccidiostats that function by inhibiting thiamine transport; amprolium is typical (Figure C-2: K). Several of these compounds could be tried against clostridia. Lowe and Potter (1980) recently reported on antibacterial fluoro analogues of thiamine and the B_1-thiazole.

There are grounds for optimism about eventual discovery of new design targets. The publications of Huhtanen and co-workers have

Pyruvate: Ferrodoxin Oxidoreductase

$$CH_3{-}CO{-}COOH + H_3PO_4 \longrightarrow CH_3{-}CO{-}OPO_3H_2 + CO_2 + H_2$$

Steps of the Phosphoroclastic Reaction

1 $\boxed{\text{pyruvate}} + \text{TPP-E} \rightleftharpoons \text{HETPP} + \boxed{CO_2}$

2 $\text{HETPP-E} + \text{ferredoxin} + \text{CoA} \rightleftharpoons \text{TPP-E} + \text{acetyl-CoA} + \text{ferredoxin} \cdot H_2$

3 $\text{ferredoxin} \cdot H_2 \overset{\text{hydrogenase}}{\rightleftharpoons} \text{ferredoxin} + \boxed{H_2}$

4 $\text{acetyl-CoA} + P_i \overset{\text{phosphotransacetylase}}{\rightleftharpoons} \boxed{\text{acetyl phosphate}} + \text{CoA}$

Steps of the phosphoroclastic reaction. Steps 1 and 2 involve the enzyme pyruvate-ferrodoxin oxidoreductase and ferrodoxin. TPP-E, thiamine pyrophosphate-containing oxidoreductase. HETPP-E, hydroxyethyl-TPP-E. Steps 3 and 4 are catalyzed by hydrogenase and phosphotransacetylase, respectively.

Pyruvate-Formate Lyase

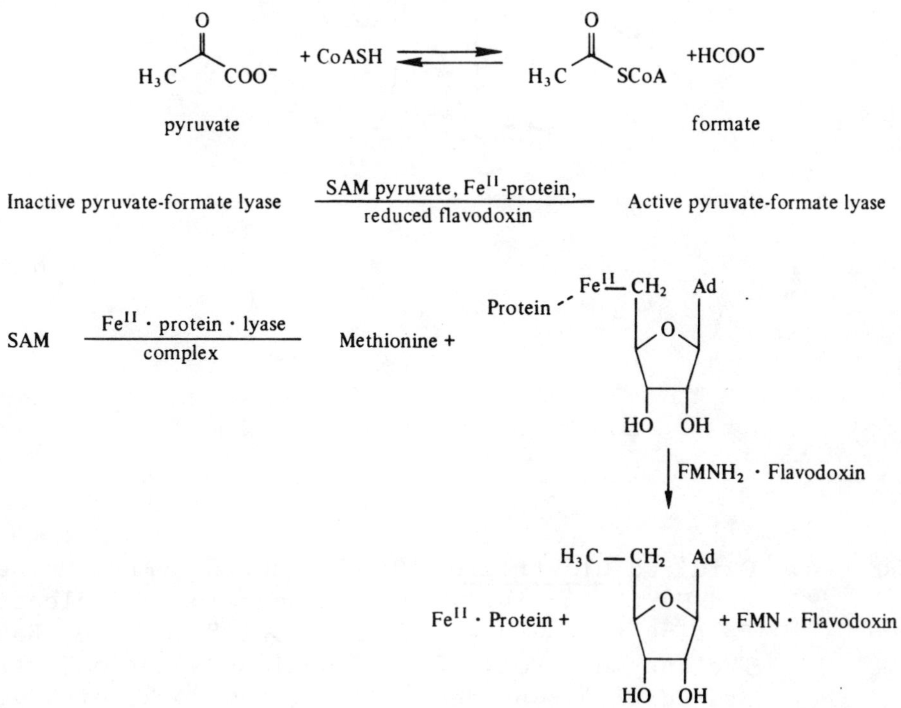

FIGURE C-3. Possible target areas in anaerobic energy metabolism. From Gottschalt, 1979 (p. 183), and Enzymatic Reaction Mechanisms by Christopher Walsh. W. H. Freeman and Company. Copyright © 1979, with permission.

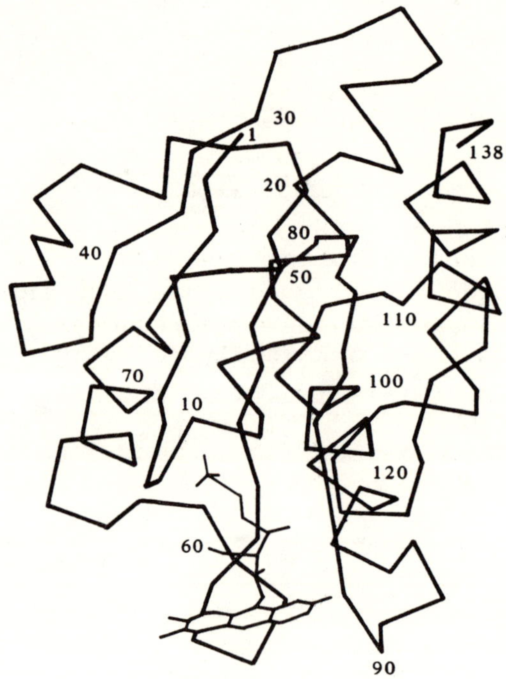

FIGURE C-4. Oxidized Clostridium MP flavodoxin. From Mayhew, S. G. and M. L. Ludwig, 1975. Flavodoxins and electron-transferring flavoproteins. Pp. 55-118 in P. D. Boyer, ed. The Enzymes, Vol. 12. Oxidation-Reduction, Part B, 3rd Edition. Academic Press, New York, with permission.

indicated that *C. botulinum* is vulnerable to a variety of agents--structures as diverse as bile acids (Huhtanen, 1979), 4-hydroxybenzoates (Dymicky and Huhtanen, 1979), aliphatic long-chain amines and aminodiamides (Huhtanen and Micich, 1978), and unknown spice extractive constituents (Huhtanen, 1980). Obviously, much further screening should be encouraged. Given effective agents, investigations of their mechanisms of action are good routes to recognition of sensitive enzymes. In the real world, the study of drugs teaches biochemistry, rather than the other way around.

A likely subject for mechanism studies may be marfanil (Figure C-2: L), the World War II gas-gangrene agent. Perhaps a Sloane degradation (Sloane, 1964, 1965; Sloane and Heinemann, 1967; Sloane *et al.*, 1963) to 4-hydroxybenzenesulfonamide occurs, and this would be related to the anticlostridial 4-hydroxybenzoates.

Other intriguing subjects for mechanism research may be the polypeptide fraction from haddock, which is claimed to be useful for prevention of clostridial outgrowth (Nickerson and Zak, 1973) and the antibiotic nisin (Rayman *et al.*, 1981).

In his interesting spice studies, Huhtanen (1980) noted that mace extractive has high activity against *C. botulinum*. A recent patent (Lion Dentifrice, 1980) covered a composition containing 2,6-dihydroxy-9-(2,5-dihydroxyphenyl)nonanophenone (Figure C-2: M), a mace constituent that is claimed to be useful for suppression of dental plaque formation ascribed to *Streptococcus mutans*. It is possible that this compound is the anticlostridial agent. That possibility should be confirmed, and other active principles of spices should also be identified. If screening is encouraged, it may be advisable to bias candidate selection in favor of natural products, such as fermentation broths and plant extractives, because there is a desirable presumption of biodegradability in active natural products.

REFERENCES

Ando, Y. 1973. Studies on germination of spores of clostridial species capable of causing food poisoning. I. Factors affecting the germination of spores of *Clostridium botulinum* type A in chemically defined medium. J. Food Hyg. Soc. Japan 14:457-461.

Dymicky, M., and C. N. Huhtanen. 1979. Inhibition of *Clostridium botulinum* by p-hydroxybenzoic acid n-alkyl esters. Antimicrob. Agents Chemother. 15:798-801.

Fushiya, S., Y. Sato, and S. Nozoe. 1980. Structure and synthesis of avenic acid B. A new amino acid possessing an iron chelating activity. Chem. Lett. 1215-1218.

Gottschalk, G. 1979. Bacterial Metabolism. Springer-Verlag, New York.

Gottschalk, G., and J. R. Andreesen. 1979. Energy metabolism in anaerobes. Internatl. Rev. Biochem. 21:85-115.

Graham, D. W., J. E. Brown, W. T. Ashton, R. D. Brown, and E. F. Rogers. 1977. Anticoccidial riboflavine antagonists. Experientia 33:1274-1275.

Hayashi, R., Z. Yoshii, T. Haroda, S. Kobayashi, D. Miura, A. Okusa, A. Takamura, Y. Sera, N. Kawaguchi, and K. Nigota. 1973. Thiaminase of the genus Clostridium. III. Thiaminase production of *Clostridium botulinum*. Vitamins 47:105-110.

Huhtanen, C. N. 1979. Bile acid inhibition of *Clostridium botulinum*. Appl. Environ. Microbiol. 38:216-218.

Huhtanen, C. N. 1980. Inhibition of *Clostridium botulinum* by spice extracts and aliphatic alcohols. J. Food Prot. 43:195-197.

Huhtanen, C. N., and T. J. Micich. 1978. Inhibition of *Clostridium botulinum* by aliphatic amines and long chain aliphatic aminodiamides. J. Am. Oil Chem. Soc. 55:854-855.

Kasai, S., R. Miura, and K. Matsui. 1975. Chemical structure and some properties of roseoflavin. Bull. Chem. Soc. Japan 48:2877-2880.

Kluger, R., and D. C. Pike. 1977. Active site generated analogues of reactive intermediates in enzymic reactions. Potent inhibition of pyruvate dehydrogenase by a phosphonate analogue of pyruvate. J. Am. Chem. Soc. 99:4504-4506.

Lion Dentifrice. 1980. Oral composition for suppression of dental plaque formation, containing 2',6'-di:hydroxy-9-(2,5-dihydroxy-phenylocyl:phenone obtained from mace. Japanese Patent 5:5104-5210.

Lowe, G., and B. V. L. Potter. 1980. Bacteriostatic activity of fluoro analogues of 5-(2-hydroxyethyl)-4-methyl thiazole, a metabolic intermediate in the biosynthesis of thiamine. J. Chem. Soc. Perkin I. 2026-2028.

Mayhew, S. G., and M. L. Ludwig. 1975. Flavodoxins and electron transferring flavoproteins. Pp. 55-188 in P. D. Boyer, ed. The Enzymes, Vol. 12. Oxidation-Reduction, Part B., 3rd Ed. Academic Press, New York.

Nickerson, J. T. R., and J. M. Zak. 1973. Treating food with polypeptide haddock extract to prevent the outgrowth of _Clostridium botulinum_. U.S. Patent 3,765,908.

Rayman, M. K., B. Aris, and A. Hurst. 1981. Nisin-A possible alternative or adjunct to nitrite in the preservation of meats. Appl. Environ. Microbiol. 41:375-380.

Raymond, K. N., and C. J. Carrano. 1979. Coordination chemistry and microbial iron transport. Accounts Chem. Res. 12:183-190.

Sloane, N. H. 1964. Microsomal hydroxymethylation of the benzene ring. An intermediate step in enzymic hydroxylation. Biochim. Biophys. Acta 81:408-410.

Sloane, N. H. 1965. Hydroxymethylation of the benzene ring. 1. Microsomal formation of phenol via prior hydroxymethylation of benzene. Biochim. Biophys. Acta 107:599-602.

Sloane, N. H., and M. Heinemann. 1967. Hydroxymethylation of the benzene ring. II. Sequential formation of p-hydroxy-aniline _via_ hydroxymethylation of aniline by microsomes. Biochim. Biophys. Acta 141:47-54.

Sloane, N. H., K. G. Untch, and A. W. Johnson. 1963. Metabolites of p-aminobenzoic acid. III. A metabolic pathway of p-amino-benzoic acid resulting in the sequential formation of p-aminobenzyl alcohol and p-hydroxyaniline. Biochim. Biophys. Acta 78:588-593.

Tompkin, R. B., L. N. Christiansen, and A. B. Shaparis. 1979. Iron and the antibotulinal efficacy of nitrite. Appl. Env. Microbiol. 37:351-353.

Walsh, C. 1979. Enzymatic Reaction Mechanisms. W. H. Freeman, San Francisco. 978 pp.

APPENDIX D

THE PROBABILITY OF OUTGROWTH AND TOXIGENESIS FROM *CLOSTRIDIUM BOTULINUM* SPORES IN TEMPERATURE-ABUSED CURED MEATS

By

A. H. W. Hauschild
Bureau of Microbial Hazard
Health and Welfare Canada
Ottawa, Ontario

The probability (p) of outgrowth and toxigenesis in a product inoculated with *C. botulinum* spores and held at simulated abuse temperatures can be calculated with the formula

$$p = [\ln n/q]/s,$$

where n is the number of packages in the test group, q is the number of nontoxic packages (Halvorson and Ziegler, 1933), and s is the number of challenge spores per package. This method assumes that the probability that a particular spore will ultimately cause toxigenesis is not affected by the presence of other spores (Hauschild *et al.*, in press). Calculation of probabilities facilitates the quantitative comparison of results of various challenge studies.

Tables D-1, D-2, and D-3 show the probabilities derived with this method for a number of challenge studies on bacon, canned perishable pork, and some other products made with different additions of nitrite. In experiments on canned perishable pork, the probabilities have been calculated for a simulated temperature-abuse period of 2 wk, as well as for 1 wk, because canned products are more likely than others to be held without correct refrigeration.

A variety of factors other than nitrite concentration, such as brine concentration and added carbohydrate, may affect p; some of these are listed in the tables. Not considered in the tables is the possible effect on p of the period of refrigeration preceding the temperature abuse. In challenge experiments, the meats are commonly incubated immediately after production, although commercial products are likely to be abused much later. Prior storage might affect outgrowth of *C. botulinum* in two opposite ways: the decrease in residual nitrite could facilitate outgrowth, and the development of a competitive microflora could retard it. Spore viability may also decrease during storage, and germinated spores whose outgrowth is prevented by the low temperature may die. Tompkin *et al.* (1978b) showed that the likelihood of outgrowth and toxigenesis during abuse of canned comminuted pork was greater after storage at 10°C for 2-18 wk than without prior storage. A marginal increase in p was noted when this product was held at 4.4°C for 10 wk before abuse (Tompkin *et al.*, 1979b). Other workers have found that p was smaller after storage at refrigeration temperatures. Pierson (1979) found that p for bacon decreased with storage at 10°C for 2-8 wk, and Hauschild *et al.* (in press) found that p was marginally smaller for liver sausage after refrigeration at 8°C for 6 wk. Hustad *et al.* (1973) found that storage of wieners for 3 wk at 7°C had little or no effect on p.

It can be seen from Table D-1 that the bulk of p values for bacon at currently permissible sodium nitrite concentrations (120 mg/kg in the United States, 150 mg/kg in Canada) ranged from 10^{-7} to 10^{-5}. The values increased by a factor of about 10 when nitrite was omitted.

The most obvious influence on p in canned perishable pork (Table D-2), other than that of added nitrite, is that of sodium isoascorbate (erythorbate) or ascorbate. The presence of isoascorbate or ascorbate at 200 mg/kg reduces p noticeably.

Table D-3 shows p for a number of other products, but some of these values are based on low numbers of replicates.

The probabilities shown in Tables D-1, D-2, and D-3 are summarized in Table 4-2 and discussed in Chapters 3 and 4.

TABLE D-1

Probability of Outgrowth and Toxigenesis from *C. botulinum* Spores in Vacuum-Packaged Bacon During Simulated Temperature Abuse

Sodium Nitrite, mg/kg	Sodium Isoascorbate, mg/kg	Polyphosphates, %	pH	Salt, %	Brine, %	Sugar, %	Temperature Abuse Days	Temperature Abuse °C	Spores/ Package	Toxic Pack./ Total Pack.	p[a]	Reference
0	230	0.26	NL[b]	1.5	4.9	0.3	7	27	6.5×10^3	4/5	2×10^{-4}	Christiansen *et al.*, 1974
									5.4×10^5	5/5	$>3 \times 10^{-6}$	
									5.0×10^3	0/2	$<1 \times 10^{-4}$	
									4.2×10^5	2/2	$>2 \times 10^{-6}$	
	0	0	6.15	1.7	5.5	0.7	8	30	2.3×10^4	0/4	$<1 \times 10^{-5}$	Collins-Thompson *et al.*, 1974
									2.3×10^6	1/4	1×10^{-7}	
									2.3×10^4	0/8	$<6 \times 10^{-6}$	
									2.3×10^6	0/8	$<6 \times 10^{-8}$	
	550	0	6.25	1.3	3.6	0.1	7	27	1.25×10^5	0/25	$<3 \times 10^{-7}$	Ivey *et al.*, 1978[c]
		0.4							1.25×10^5	0/25	$<3 \times 10^{-7}$	
	550	0.4	6.5	1.3	3.5	0.1	7	27	8.9×10^4	4/4	$>2 \times 10^{-5}$	Pierson, 1978
									3.2×10^4	2/4	2×10^{-5}	
		0.2	6.25	1.3	3.7	0.1	7	27	2.2×10^4	0/4	$<1 \times 10^{-5}$	
									3.0×10^4	0/4	$<1 \times 10^{-5}$	
	550	0.3	NL	1.14	3.3	0.1	7	27	3.2×10^4	29/30	1×10^{-4}	Pierson, 1979[d]
										21/30	4×10^{-5}	
										15.5/30[e]	2×10^{-5}	
										8.25/30	1×10^{-5}	
	550	0.3	∿6.5	1.5	NL	0.06	7	27	2.0×10^5	99/100	2×10^{-5}	U.S. Department of Agriculture, 1979
		0.08	∿6.2	1.7	NL	0	7	27	2.0×10^5	31/100	2×10^{-6}	
		0.12	∿6.4	1.8	NL	0.12	7	27	2.0×10^5	28/99	2×10^{-6}	
		0.46	∿6.4	1.4	NL	>0.5	7	27	2.0×10^5	20/100	1×10^{-6}	
	550	0.31	6.2	1.5	NL	0	7	27	1.0×10^5	13/13	$>3 \times 10^{-5}$	Tanaka *et al.*, 1980
						0.5	7	27	1.0×10^5	14/16	2×10^{-5}	
	550	0.5	NL	1.5	4.5	0.8	10	27	1.4×10^5	7/10	9×10^{-6}	Sofos *et al.*, 1980

TABLE D-1 (continued)

Sodium Nitrite, mg/kg	Sodium Isoascorbate, mg/kg	Poly-phosphates, %	pH	Salt, %	Brine, %	Sugar, %	Temperature Abuse Days	Temperature Abuse °C	Spores/ Package	Toxic Pack./ Total Pack.	p[a]	Reference
	550	0.3	6.05	1.5	4.3	0.25	9	27	2.0×10^{2}	18/20	1×10^{-2}	D. R. Rowley, personal communication[f]
									1.6×10^{4}	16/19	1×10^{-4}	
			6.1	1.5	4.0	0.75	9	27	2.8×10^{5}	7/20	2×10^{-6}	
	550	0.3	6.55	1.7	5.0	0.11	7	27	3.7×10^{4}	1/4	8×10^{-6}	Pierson _et al._, 1981
	550	0.3	6.25	1.8	4.7	0.11	7	27	1.9×10^{5}	16/19	1×10^{-5}	M. D. Pierson, personal communication
20	550	0.3	6.1	1.5	4.0	0.75	9	27	2.8×10^{5}	13/20	4×10^{-6}	D. R. Rowley, personal communication[f]
30	230	0.26	NL	1.5	4.9	0.3	7	27	6.5×10^{3}	4/5	2×10^{-4}	Christiansen _et al._, 1974
									5.4×10^{5}	1/5	4×10^{-7}	
									5.0×10^{3}	0/2	$<1 \times 10^{-4}$	
									4.2×10^{5}	2/2	$>2 \times 10^{-6}$	
40	550	0	6.25	1.3	3.6	0.1	7	27	1.25×10^{5}	0/25	$<3 \times 10^{-7}$	Ivey _et al._, 1978[c]
		0.2							1.25×10^{5}	0/25	$<3 \times 10^{-7}$	
		0.4							1.25×10^{5}	0/25	$<3 \times 10^{-7}$	
	550	0.4	6.5	1.3	3.5	0.1	7	27	8.9×10^{4}	0/4	$<3 \times 10^{-6}$	Pierson, 1978
									3.2×10^{4}	0/4	$<9 \times 10^{-6}$	
		0.2	6.25	1.3	3.7	0.1	7	27	2.2×10^{4}	0/4	$<1 \times 10^{-5}$	
									3.0×10^{4}	0/4	$<1 \times 10^{-5}$	
	550	0.31	6.2	1.5	NL	0	7	27	1.0×10^{5}	22/25	2×10^{-5}	Tanaka _et al._, 1980
						>0.5	7	27	1.0×10^{5}	3/3	$>1 \times 10^{-5}$	
	550	0.3	6.05	1.5	4.3	0.25	9	27	2.0×10^{2}	2/20	5×10^{-4}	D. R. Rowley, personal communication[f]
									1.6×10^{4}	13/19	7×10^{-5}	
	550	0.3	6.55	1.7	5.0	0.11	7	27	3.7×10^{4}	0/4	$<8 \times 10^{-6}$	Pierson _et al._, 1981
	550	0.3	6.25	1.8	4.7	0.11	7	27	1.9×10^{5}	1/4	2×10^{-6}	M. D. Pierson, personal communication

50	0	0	6.15	1.7	5.5	0.7	8	30	2.3×10^4 2.3×10^6 2.3×10^4 2.3×10^6	1/4 1/4 0/8 1/8	1×10^{-5} 1×10^{-7} $<6 \times 10^{-6}$ 6×10^{-8}	Collins-Thompson <u>et</u> <u>al</u>., 1974
60	230	0.26	NL	1.5	4.9	0.3	7	27	6.5×10^3 5.4×10^5 5.0×10^3 4.2×10^5	0/5 0/5 0/2 0/2	$<3 \times 10^{-5}$ $<4 \times 10^{-7}$ $<1 \times 10^{-4}$ $<2 \times 10^{-6}$	Christiansen <u>et</u> <u>al</u>., 1974
	550	0.31	6.2	1.5	NL	$\geq$0.5	7	27	1.0×10^5	0/10	$<1 \times 10^{-6}$	Tanaka <u>et</u> <u>al</u>., 1980
80	550	0	6.25	1.3	3.6	0.1	7	27	1.25×10^5	0/25	$<3 \times 10^{-7}$	Ivey et al., 1978[c]
	550	0.31	6.2	1.5	NL	$\geq$0.5	7	27	1.0×10^5	2/10	2×10^{-6}	Tanaka <u>et</u> <u>al</u>., 1980
100	0	0	6.15	1.7	5.5	0.7	8	30	2.3×10^4 2.3×10^6 2.3×10^4 2.3×10^6	0/4 2/4 0/8 2/8	$<1 \times 10^{-5}$ 3×10^{-7} $<6 \times 10^{-6}$ 1×10^{-7}	Collins-Thompson <u>et</u> <u>al</u>., 1974
120	230	0.26	NL	1.5	4.9	0.3	7	27	6.5×10^3 5.4×10^5 5.0×10^3 4.2×10^5	0/5 0/5 0/2 0/2	$<3 \times 10^{-5}$ $<4 \times 10^{-7}$ $<1 \times 10^{-4}$ $<2 \times 10^{-6}$	Christiansen <u>et</u> <u>al</u>., 1974
	550	0	6.25	1.3	3.6	0.1	7	27	1.25×10^5	0/25	$<3 \times 10^{-7}$	Ivey <u>et</u> <u>al</u>., 1978[c]
	550	0.4	6.5	1.3	3.5	0.1	7	27	8.9×10^4 3.2×10^4	0/4 0/4	$<3 \times 10^{-6}$ $<9 \times 10^{-6}$	Pierson, 1978
		0.2	6.25	1.3	3.7	0.1	7	27	2.2×10^4 3.2×10^4	0/4 0/4	$<3 \times 10^{-5}$ $<1 \times 10^{-5}$	
	550	0.3	NL	1.14	3.3	0.1	7	27	3.2×10^4	21/30 0/30 0/30 0/30	4×10^{-5} $<1 \times 10^{-6}$ $<1 \times 10^{-6}$ $<1 \times 10^{-6}$	Pierson, 1979[d]

TABLE D-1 (continued)

Sodium Nitrite, mg/kg	Sodium Isoascorbate, mg/kg	Polyphosphates, %	pH	Salt, %	Brine, %	Sugar, %	Temperature Abuse Days	Temperature Abuse °C	Spores/ Package	Toxic Pack./ Total Pack.	p[a]	Reference
	550	0.3	6.5	1.5	NL	0.06	7	27	2.0×10^5	88/200	3×10^{-6}	U.S. Department of Agriculture, 1979
		0.08	6.2	1.7	NL	0	7	27	2.0×10^5	20/194	5×10^{-7}	
		0.12	6.4	1.8	NL	0.12	7	27	2.0×10^5	3/200	8×10^{-8}	
		0.46	6.4	1.4	NL	0.5	7	27	2.0×10^5	0/200	$<2 \times 10^{-8}$	
	550	0.31	6.2	1.5	NL	0	7	27	1.0×10^5	4/16	3×10^{-6}	Tanaka et al., 1980
						≥ 0.5	7	27	1.0×10^5	1/36	3×10^{-7}	
	550	0.5	NL	1.8	4.9	0.8	10	27	1.4×10^5	0/10	$<8 \times 10^{-7}$	Sofos et al., 1980
	550	0.3	6.05	1.5	4.3	0.25	9	27	2.0×10^2	1/15	3×10^{-4}	D. R. Rowley, personal communication[f]
									1.6×10^4	7/15	4×10^{-5}	
			6.1	1.5	4.0	0.75	9	27	2.8×10^5	0/20	$<2 \times 10^7$	
	550	0.3	6.55	1.7	5.0	0.11	7	27	3.7×10^4	0/4	$<8 \times 10^{-6}$	Pierson et al., 1981
	550	0.3	6.25	1.8	4.7	0.11	7	27	1.9×10^5	0/4	$<2 \times 10^{-6}$	M. D. Pierson, personal communication
150	0	0	6.15	1.7	5.5	0.7	8	30	2.3×10^4	1/4	1×10^{-5}	Collins-Thompson et al., 1974
									2.3×10^6	2/4	3×10^{-7}	
									2.3×10^4	0/8	$<6 \times 10^{-6}$	
									2.3×10^6	0/8	$<6 \times 10^{-8}$	

[a] $p = (\ln n/q)/s$, where n = no. packages, q = no. nontoxic packages, and s = no. challenge spores/package.
[b] NL = not listed.
[c] Only swollen packages were examined for toxin.
[d] Before abuse at 27°C, the four groups of packages were held at 10 ° C for 0, 2, 5, and 8 wk, respectively.
[e] Values for nonswollen packages were extrapolated from X/4.
[f] Swollen packages recorded after 7 d and assayed for toxin after 9 d.

TABLE D-2

Probability of Toxin or Gas Production from *C. botulinum* Spores In Canned Comminuted Pork During Simulated Temperature Abuse[a,b]

Sodium Nitrite, mg/kg	Sodium Isoascorbate,[d] 200 mg/kg	Spores/ Can	Temperature Abuse[c] 1 Week at 27 °C Toxic Cans/ Total Cans	1 Week at 27 °C p[e]	2 Weeks at 27 °C Toxic Cans/ Total Cans	2 Weeks at 27 °C p[e]	Reference
0	-	6.2×10^3	25/25	$>5 \times 10^{-4}$	--	--	Tompkin et al., 1978a
		1.1×10^3	24/25	3×10^{-3}	25/25	$>3 \times 10^{-3}$	
50		6.2×10^3	0/25	$<7 \times 10^{-6}$	25/25	$>5 \times 10^{-4}$	Tompkin et al., 1978a
		1.1×10^3	1/25	4×10^{-5}	23/25	2×10^{-3}	
			22/25	2×10^{-3}	25/25	$>3 \times 10^{-3}$	
			12/25	6×10^{-4}	25/25	$>3 \times 10^{-3}$	
		1.3×10^3	17/25	9×10^{-4}	25/25	$>2 \times 10^{-3}$	
		8.0×10^3[f]	0/25	$<5 \times 10^{-6}$	25/25	$>4 \times 10^{-4}$	Tompkin et al., 1978c
156		1.3×10^3	0/25	$<3 \times 10^{-5}$	12/25	5×10^{-4}	Tompkin et al., 1978a
		7.7×10^3	0/25	$<5 \times 10^{-6}$	0/25	$<5 \times 10^{-6}$	Tompkin et al., 1979a
0	+	4.0×10^5	30/40	3×10^{-6}	34/40	5×10^{-6}	Christiansen et al., 1973
		1.0×10^4	25/25	$>3 \times 10^{-4}$	--	--	Tompkin et al., 1977
		1.5×10^4	25/25	$>2 \times 10^{-4}$	--	--	
		1.0×10^4	25/25[g]	$>3 \times 10^{-4}$	--	--	
		4.2×10^3	25/25[g]	$>8 \times 10^{-4}$	--	--	
		1.7×10^4	25/25[g]	$>2 \times 10^{-4}$	--	--	
		6.2×10^3	25/25	$>5 \times 10^{-4}$	--	--	Tompkin et al., 1978a
		1.1×10^3	23/25	2×10^{-3}	25/25	$>3 \times 10^{-3}$	
		1.0×10^4	25/25	$>3 \times 10^{-4}$	--	--	Ivey and Robach, 1978[h]
		1.5×10^4	0/25	$>3 \times 10^{-6}$	25/25	$>2 \times 10^{-4}$	
50	+	7.2×10^3	2/40	7×10^{-6}	20/40	1×10^{-4}	Christiansen et al., 1973
		1.0×10^4	0/25	$<4 \times 10^{-6}$	6/25	3×10^{-5}	Tompkin et al., 1977
		1.5×10^4	--	--	0/25	$<3 \times 10^{-6}$	
		4.2×10^3	--	--	0/25	$<1 \times 10^{-5}$	
		1.7×10^4	--	--	0/25	$<2 \times 10^{-6}$	
		2.3×10^3	--	--	0/25	$<2 \times 10^{-5}$	
		5.2×10^3	--	--	0/25	$<8 \times 10^{-6}$	
		6.2×10^3	--	--	0/25	$<7 \times 10^{-6}$	Tompkin et al., 1978a
		1.1×10^3	--	--	0/25	$<4 \times 10^{-5}$	
			0/25	$<4 \times 10^{-5}$	1/25	4×10^{-5}	
			0/25	$<4 \times 10^{-5}$	5/25	2×10^{-4}	
			0/25	$<3 \times 10^{-5}$	16/25	8×10^{-4}	
		8.0×10^3[f]	--	--	0/25	$<5 \times 10^{-6}$	Tompkin et al., 1978c
		1.0×10^4	--	--	0/25	$<4 \times 10^{-6}$	Ivey and Robach, 1978
		1.5×10^4	--	--	0/100	$<7 \times 10^{-7}$	
100	+	4.0×10^5	0/40	$<6 \times 10^{-8}$	9/40	6×10^{-7}	Christiansen et al., 1973
		1.0×10^4	--	--	0/25	$<4 \times 10^{-6}$	Tompkin et al., 1977
		1.5×10^4	--	--	0/25	$<3 \times 10^{-6}$	
		4.2×10^3	--	--	0/25	$<1 \times 10^{-5}$	
		1.7×10^4	--	--	0/25	$<2 \times 10^{-6}$	
		2.3×10^3	--	--	0/25	$<2 \times 10^{-5}$	
		5.2×10^3	--	--	0/25	$<8 \times 10^{-6}$	
		1.0×10^4	--	--	0/25	$<4 \times 10^{-6}$	Ivey and Robach, 1978
		1.5×10^4	--	--	0/25	$<3 \times 10^{-6}$	

TABLE D-2 (continued)

Sodium Nitrite, mg/kg	Sodium Isoascorbate,[d] 200 mg/kg	Spores/ Can	Temperature Abuse[c]: 1 Week at 27°C, Toxic Cans/ Total Cans	1 Week at 27°C, p[e]	2 Weeks at 27°C, Toxic Cans/ Total Cans	2 Weeks at 27°C, p[e]	Reference
150 or 156	+	7.2×10^3	--	--	0/40	$<4 \times 10^{-6}$	Christiansen et al., 1973
		1.0×10^4	--	--	0/25	$<4 \times 10^{-6}$	Tompkin et al., 1977
		1.5×10^4	--	--	0/25	$<3 \times 10^{-6}$	
		4.2×10^3	--	--	0/25	$<1 \times 10^{-5}$	
		1.7×10^4	--	--	0/25	$<2 \times 10^{-6}$	
		2.3×10^3	--	--	0/25	$<2 \times 10^{-5}$	
		5.2×10^3	--	--	0/25	$<8 \times 10^{-6}$	
		1.3×10^3	--	--	0/25	$<3 \times 10^{-5}$	Tompkin et al., 1978a
		8.0×10^3[f]	--	--	0/25	$<5 \times 10^{-6}$	Tompkin et al., 1978b
		1.0×10^4	--	--	0/25	$<4 \times 10^{-6}$	Ivey and Robach, 1978
		1.5×10^4	--	--	0/25	$<3 \times 10^{-6}$	
		7.7×10^3	--	--	0/25	$<5 \times 10^{-6}$	Tompkin et al., 1979a
		8.0×10^3	--	--	0/50	$<3 \times 10^{-6}$	Tompkin et al., 1979b

[a]Toxigenesis listed by Christiansen et al., 1973; swells (generally confirmed by toxin assays) listed in other studies.

[b]Where recorded (Christiansen et al., 1973; Ivey and Robach, 1978; Tompkin et al., 1977, 1978a): ph, 5.8-6.3; brine, 3.7-4.1%; sugar, 0.5%.

[c]Identical results after 1 and 2 wk entered only once.

[d]Isoascorbate added (+) or not added (-).

[e]$p = (\ln n/q)/s$, where n = no. cans, q = no. nontoxic nonswollen cans, and s = no. challenge spores/can.

[f]Target input of spores.

[g]After 8 d of incubation.

[h]Tests 1 and 2 listed under Tompkin et al., 1977.

TABLE D-3

Probability of Outgrowth and Toxigenesis from C. botulinum Spores in Vacuum-Packaged Meats and Liver Sausage During Simulated Temperature Abuse

					Temperature Abuse[a]				
					1 Week at 27°C		2 Weeks at 27°C		
Product	Sodium Nitrite, mg/kg	Brine, %	Sugar, %	Spores/ Package	Toxic Pack./ Total Pack.	p[b]	Toxic Pack./ Total Pack.	p[b]	Reference
Frankfurters	0	4.5	2.8[c]	1.5×10^4	0/5	$<1 \times 10^{-5}$	5/5	$>1 \times 10^{-4}$	Hustad *et al.*, 1973
					0/5	$<1 \times 10^{-5}$	4/5	1×10^{-4}	
		4.8	2.8[c]	$\sim 2 \times 10^4$	0/5	$<1 \times 10^{-5}$	4/5	8×10^{-5}	Bowen *et al.*, 1974
					1/5	1×10^{-5}	4/5	8×10^{-5}	
					1/5	1×10^{-5}	5/5	$>8 \times 10^{-5}$	
	30	4.8	2.8	$\sim 2 \times 10^4$	0/5	$<1 \times 10^{-5}$	2/5	3×10^{-5}	Bowen *et al.*, 1974
					3/5	5×10^{-5}	5/5	$>8 \times 10^{-5}$	
					3/5	5×10^{-5}	5/5	$>8 \times 10^{-5}$	
	50	4.5	2.8	1.5×10^4	0/5	$<1 \times 10^{-5}$	0/5	$<1 \times 10^{-5}$	Hustad *et al.*, 1973
					0/5	$<1 \times 10^{-5}$	0/5	$<1 \times 10^{-5}$	
		4.8	2.8	$\sim 2 \times 10^4$	0/5	$<1 \times 10^{-5}$	4/5	8×10^{-5}	Bowen *et al.*, 1974
					1/5	1×10^{-5}	4/5	8×10^{-5}	
					0/5	$<1 \times 10^{-5}$	1/5	1×10^{-5}	
	100 or 156	4.8	2.8	$\sim 2 \times 10^4$	0/5	$<1 \times 10^{-5}$	0/5	$<1 \times 10^{-5}$	
					0/5	$<1 \times 10^{-5}$	0/5	$<1 \times 10^{-5}$	
					0/5	$<1 \times 10^{-5}$	0/5	$<1 \times 10^{-5}$	
Chicken frank-furters	0	3.2-3.4	3.0[d]	4.2×10^2	5/5	$>4 \times 10^{-3}$	--	--	Christiansen *et al.*, 1977
	50		3.0		5/5	$>4 \times 10^{-3}$	--	--	
	100		0		0/5	$<5 \times 10^{-4}$	1/5	5×10^{-4}	
			3.0		0/5	$<5 \times 10^{-4}$	4/5	4×10^{-3}	
	156		3.0		0/5	$<5 \times 10^{-4}$	4/5	4×10^{-3}	
Turkey ham	0	2.7-3.4	0	1.2×10^3	5/5	$>1 \times 10^{-3}$	--	--	Christiansen *et al.*, 1977
	50				4/5[e]	1×10^{-3}	--	--	
	100				1/5	2×10^{-4}	5/5	$>1 \times 10^{-3}$	
	156				0/5	$<2 \times 10^{-4}$	5/5	$>1 \times 10^{-3}$	

TABLE D-3 (continued)

Product	Sodium Nitrite, mg/kg	Brine, %	Sugar, %	Spores/ Package	Temperature Abuse[a] 1 Week at 27°C Toxic Pack./ Total Pack.	p^b	2 Weeks at 27°C Toxic Pack./ Total Pack.	p^b	Reference
Fermented sausages	0[f] 0[h]	NL[g]	2.0	4.6×10^4	1/3 0/3	9×10^{-6} $<9 \times 10^{-6}$	0/3 0/3	$<9 \times 10^{-6}$ $<9 \times 10^{-6}$	Christiansen et al., 1975
	50,100, or 150[f]				0/3	$<9 \times 10^{-6}$	0/3	$<9 \times 10^{-6}$	
	50 or 150[f]				0/3	$<9 \times 10^{-6}$	0/3	$<9 \times 10^{-6}$	
Liver sausages	0	NL	0	$\sim 10^5$	3/3	$>1 \times 10^{-5}$	--	--	Ala-Huikku et al., 1977[i]
	100				0/3	$<4 \times 10^{-6}$	0/3	$<4 \times 10^{-6}$	
	0	5.4	0	8.3×10^4	0/5	$<3 \times 10^{-6}$	5/5	$>2 \times 10^{-5}$	Hauschild et al., in press
	50,100, or 150				0/5	$<3 \times 10^{-6}$	0/5	$<3 \times 10^{-6}$	
	0	3.8-4.2		8.3x $(10^1-10^4)^j$		3×10^{-4} 3×10^{-4} 1×10^{-2}		$>2 \times 10^{-3}$ 1×10^{-3}	Hauschild et al., in press
	50					8×10^{-5} 8×10^{-3} 1×10^{-2}		$>2 \times 10^{-3}$ $>2 \times 10^{-2}$	
	100					1×10^{-5} 2×10^{-4} 2×10^{-2}		$>2 \times 10^{-3}$ 6×10^{-3} --	
	150					$<2 \times 10^{-6}$ $<2 \times 10^{-6}$ $<2 \times 10^{-6}$		2×10^{-6} 1×10^{-4} 1×10^{-3}	

[a] Identical results after 1 and 2 wk entered only once.
[b] $p = (\ln n/q)/s$, where n = no. packages, q = no. nontoxic packages, and s = no. challenge spores/package.
[c] Including 1.8% corn syrup solids.
[d] Including 2.0% corn syrup solids.
[e] Toxicities after 6 d.
[f] Without starter culture.
[g] NL, not listed
[h] With starter culture.
[i] Liver content only 17% of meat block; incubation temp., 25°C.
[j] Multidose spore challenge.

REFERENCES

Ala-Huikku, K., E. Nurmi, H. Pajulahti, and M. Raevouri. 1977. Effect of nitrite, storage temperature and time on *Clostridium botulinum* type A toxin formation in liver sausage. Eur. J. Appl. Microbiol. 4:145-149.

Bowen, V. G., J. G. Cerveny, and R. H. Deibel. 1974. Effect of sodium ascorbate and sodium nitrite on toxin formation of *Clostridium botulinum* in wieners. Appl. Microbiol. 27:605-606.

Christiansen, L. N., R. W. Johnston, D. A. Kautter, J. W. Howard, and W. J. Aunan. 1973. Effect of nitrite and nitrate on toxin production by *Clostridium botulinum* and on nitrosamine formation in perishable canned comminuted cured meat. Appl. Microbiol. 25: 357-362 (Erratum 26:653).

Christiansen, L. N., R. B. Tompkin, A. B. Shaparis, T. V. Kueper, R. W. Johnston, D. A. Kautter, and O. J. Kolari. 1974. Effect of sodium nitrite on toxin production by *Clostridium botulinum* in bacon. Appl. Microbiol. 27:733-737.

Christiansen, L. N., R. B. Tompkin, A. B. Shaparis, R. W. Johnston, and D. A. Kautter. 1975. Effect of sodium nitrite and nitrate on *Clostridium botulinum* growth and toxin production in a summer style sausage. J. Food Sci. 40:488-490.

Christiansen, L. N., J. C. Manuel, and L. S. Divilbiss. 1977. Effect of sodium nitrite on cured poultry products. Special Poultry Research Committee, Response to Federal Drug Administration Notice Published September 2, 1977 (Fed. Reg. 42:44376). Docket No. 77N-0222.

Collins-Thompson, D. L., P. C. Chang, C. M. Davidson, E. Larmond, and H. Pivnick. 1974. Effect of nitrite and storage temperature on the organoleptic quality and toxinogenesis by *Clostridium botulinum* in vacuum-packaged side bacon. J. Food Sci. 39:607-609.

Halvorson, H. O., and N. R. Ziegler. 1933. Application of statistics to problems in bacteriology. J. Bacteriol. 25:101-121.

Hauschild, A. H. W., R. Hilsheimer, G. Jarvis, and D. P. Raymond. In press. Contribution of nitrite to the control of *Clostridium botulinum* in liver sausage. J. Food Prot.

Hustad, G. O., J. G. Cerveny, H. Trenk, R. H. Deibel, D. A. Kautter, T. Fazio, R. W. Johnston, and O. E. Kolari. 1973. Effect of sodium nitrite and sodium nitrate on botulinal toxin production and nitrosamine formation in wieners. Appl. Microbiol. 26:22-26.

Ivey, F. J., and M. C. Robach. 1978. Effect of sorbic acid and sodium nitrite on *Clostridium botulinum* outgrowth and toxin production in canned comminuted pork. J. Food Sci. 43:1782-1785.

Ivey F. J., K. J. Shaver, L. N. Christiansen, and R. B. Tompkin. 1978. Effect of potassium sorbate on toxinogenesis by *Clostridium botulinum* in bacon. J. Food Prot. 41:621-625.

Pierson, M. D. 1978. The evaluation of potassium sorbate as an inhibitor of *Clostridium botulinum* growth and toxin production in bacon. Report submitted to Monsanto Co., St. Louis, Missouri.

Pierson, M. D. 1979. The effect of delayed temperature abuse on *Clostridium botulinum* inhibition by nitrite and sorbate in bacon. Final Report of March 14th, 1978, submitted to Monsanto Co., St. Louis, Missouri.

Pierson, M. D., K. M. Rice, and J. F. Jadlocki. 1981. Sodium Hypophosphite inhibition of *Clostridium botulinum* in bacon. Pp. 651-654 in Proceedings of the 27th European Meeting of Meat Research Workers, Vol. 2, August 26-28, 1981, Vienna, Austria.

Sofos, J. N., F. F. Busta, K. Bhothipaksa, C. E. Allen, M. C. Robach, and M. W. Paquette. 1980. Effects of various concentrations of sodium nitrite and potassium sorbate on *Clostridium botulinum* toxin production in commercially prepared bacon. J. Food Sci. 45:1285-1292.

Tanaka, N., E. Traisman, M. H. Lee, R. G. Cassens, and E. M. Foster. 1980. Inhibition of botulinum toxin formation in bacon by acid development. J. Food Prot. 43:450-457.

Tompkin, R. B., L. N. Christiansen, and A. B. Shaparis. 1977. Variation in inhibition of *C. botulinum* by nitrite in perishable canned comminuted cured meat. J. Food Sci. 42:1046-1048.

Tompkin, R. B., L. N. Christiansen, and A. B. Shaparis. 1978a. Enhancing nitrite inhibition of *Clostridium botulinum* with isoascorbate in perishable canned cured meat. Appl. Env. Microbiol. 35:59-61.

Tompkin, R. B., L. N. Christiansen, and A. B. Shaparis. 1978b. Effect of prior refrigeration on botulinal outgrowth in perishable canned cured meat when temperature abused. Appl. Env. Microbiol. 35:863-866.

Tompkin, R. B., L. N. Christiansen, and A. B. Shaparis. 1978c. Antibotulinal role of isoascorbate in cured meat. J. Food Sci. 43:1368-1370.

Tompkin, R. B., L. N. Christiansen, and A. B. Shaparis. 1979a. Iron and the antibotulinal efficacy of nitrite. Appl. Env. Microbiol. 37:351-353.

Tompkin, R. B., L. N. Christiansen, and A. B. Shaparis. 1979b. Isoascorbate level and botulinal inhibition in perishable canned cured meat. J. Food Sci. 44:1147-1149.

U.S. Department of Agriculture. 1979. Food Safety and Quality Service Four Plant Study to Investigate the Use of 40 ppm Sodium Nitrite and 0.26 Percent Potassium Sorbate in Bacon. Food Safety and Quality Service, U.S. Department of Agriculture, Washington, D.C. 54 pp.